T0236155

Ernst Franzek

Helmut Beckmann

Psychoses of the Schizophrenic Spectrum in Twins

A Discussion on the Nature – Nurture Debate in the Etiology of "Endogenous" Psychoses

SpringerWienNewYork

Priv. Doz. Dr. Ernst Franzek
Psychiatrische Klinik und Poliklinik, Universitäts-Nervenklinik,
Würzburg, Germany

Univ.-Prof. Dr. Dr. h.c. Helmut Beckmann
Psychiatrische Klinik und Poliklinik, Universitäts-Nervenklinik,
Würzburg, Germany

Translated by Susanne Bachmann-Lauer

© 1999 Springer-Verlag/Wien
Printed in Austria

Typesetting: Composition & Design Services, Minsk 220027, Belarus
Printing and binding: Manz, A-1050 Wien
Graphic design: Ecke Bonk

Printed on acid-free and chlorine-free bleached paper
SPIN: 10719091

With 2 Figures

Library of Congress Cataloging-in-Publication Data
Franzek, Ernst, 1953-
Psychoses of the schizophrenic spectrum in twins : a discussion of the
    nature-nurture debate in the etiology of "endogenous" psychoses /
    Ernst Franzek, Helmut Beckmann.
    p. cm.
Includes bibliographical references and index.
ISBN 3-211-83298-X
    1. Schizophrenia. 2. Diseases in twins. 3. Psychoses–Etiology.
I. Beckmann, H. (Helmut), Dr. II. Title.
RC514.F686 1999   99-25477
616.89'82–dc21   CIP

ISBN 3-211-83298-X Springer-Verlag Wien New York

Foreword

Modern brain research has progressed tremendously in the last decades, largely due to the development of methods in neuro-imagery, biochemistry, pharmacology and histology. Despite this, it has not yet been possible to successfully apply these advances to a scientifically based, empirically verifiable psychiatry. This primarily concerns the endogenous psychoses, to use the words of the German psychiatrist Kurt Kolle from the "Delphian Oracle".

One of the reasons for the lack of progress in psychiatric research is the inconsistent and unreliable diagnostic methods which are currently carried out through expert consensus and which are then pronounced as setting a worldwide standard. The claim of reliability among various researchers is always emphasized, although often by sacrificing an adequate level of validity. These classification systems stand in contrast to the classification of endogenous psychoses based on a clinical-empirical approach and on highly differentiated descriptions of the illness, for example according to Leonhard. In contrast to operationalized diagnostics, a certain diagnosis can be provided here only when all, and especially when the characteristic symptoms of a clinical picture are clearly present. The endogenous psychoses with so-called "schizophrenic symptomatology" are subdivided according to the Leonhard classification into three large groups of illnesses, which in turn cover distinct clinical pictures: cycloid psychoses, unsystematic and systematic schizophrenia. A series of new research results point to the nosological autonomy of these groups of illnesses.

A finding which has not yet been acknowledged internationally has prompted us at this time to take on the challenge of conducting a systematic twin study, and the great effort associated with it, while taking the Leonhard classification into account. Leonhard reported that he had not seen a single monozygotic twin with systematic schizophrenia among the large number of endogenous psychotic twin test subjects that he had observed during his life-

time. Among the dizygotic twins, however, this disorder occurred
in the frequency that was to be expected statistically. He posed the
hypothesis that the close interpersonal contact generally experienced by monozygotic twins – if they grew up together – could
possibly prevent these severe, irreversible illnesses; and that in
contrast, a lack of communication in developmental stages which
were sensitive with regard to the human psyche could predestine
the occurrence of these illnesses. Let us say at this point that we
were also unable to disprove Leonhard's findings, despite our systematic ascertainment of index-twins. We view this as a challenge
in schizophrenia research and hope that dogmatic and ideological
reservations about the classification of endogenous psychoses by
Kleist and Leonhard will be dropped so that serious scientific
discussion can begin. The spectrum of psychoses with schizophrenic and schizophrenia-like symptoms does not appear to be
a continuum of disorders, but seems rather to consist of various
subgroups with extremely different genetic, somatic and psycho-
social origins. This twin study is further proof that innovative discoveries will be attainable in the area of endogenous psychoses
only through a precise clinical-psychopathological differentiation
of the psychiatric clinical pictures and specific scientific investigation.

Würzburg, 1999   *Ernst Franzek, Helmut Beckmann*

Table of contents

Introduction

The diagnostic dilemma in psychiatry

The history of psychiatry has been marked from the very beginning by the "conflict" between leading representatives of the field concerning diagnostic interpretations. For example, Heinroth (1773–1843) identified 48 different diagnoses of mental diseases, while Neumann (1814–1884) in his textbook of psychiatry wrote: "There is only one type of mental disease. We call it insanity." For a long time there was no agreement on a systematic classification of mental disorders. Using Kahlbaum's (1828–1899) fundamental idea of clinically oriented research methodology, Kraepelin (1856–1926) finally created a diagnostic-nosological classification system, in which he divided endogenous psychoses into two large groupings: manic-depressive disease and dementia praecox. This dichotomy is based primarily on the different prognoses of the two groups: favorable prognosis in the case of manic-depressive disease and unfavorable prognosis in the case of dementia praecox. Kraepelin defined the term "manic-depressive disease" rather broadly. He also did not see a disease entity in the actual sense, but rather "a group of disorders stemming from a common root, with gradual transitions between the individual types" (Kraepelin 1909). Kraepelin's paradigm of a dichotomy of endogenous psychoses dominated psychiatric research for decades. With the introduction of the term "schizophrenia" by Bleuler (1857–1939), however, the diagnostic-prognostic dichotomy was lost. Bleuler's "group of schizophrenia" included both favorable as well as unfavorable psychoses: "... it soon became apparent, however, that many disorders which cannot be distinguished in their psychopathological description from those psychoses which lead to 'dementia' have a good prognosis, similar to manic-depressive 'insanity'. A term had to be created which encompassed illnesses with similar symptomatology, even when they result partly in recovery, partly in deficiency, and partly in 'dementia'" (Bleuler 1911). The resulting consequences for psychiatric research were summarized resignedly by Gruhle (1880–1958) in 1932: "It is somewhat discouraging to see that the controversies which took place between the years 1800 and 1850 were repeated almost identically from 1900 to 1930; the only difference being that these debates were no longer concerned with insanity as such, but rather with the three endogenous psychoses (idiopathic epilepsy, manic-depressive insanity, schizophrenia) and particularly its primary subject, schizophrenia".

In the last several years attempts have been made to solve this dilemma by means of atheoretical, operationalized diagnostics. Classifications are to be made exclusively on the basis of syndromes and without involving any kind of theoretical ideas about illness. This is also expressed by the fact that the term 'disease' was replaced for the most part by the term 'disorder' in the "Diagnostic and Statistical Manual of Mental Disorders, Third Edition, Revised (DSM-III-R) and Fourth Edition (DSM IV)" of the American Psychiatric Association, as well as in the "Tenth Revision of the International Classification of Diseases (ICD 10)" of the World Health Organization. The current state of schizophrenia research has been summarized pessimistically by Parnas (1990), similarly to Gruhle: "We have now at our disposal powerful genetic, biochemical, and brain-imaging technology. Nevertheless, there is an increased gap between these developments and the growth in our understanding of the etiology of schizophrenia.".

Independently from this development, Leonhard (1904–1988) led clinical-empirical research in a new direction. Starting from the physiological-psychological viewpoint set out by Wernicke (1848–1905) and the neuropathological-psychopathological viewpoint of Kleist (1879–1960), he integrated Kraepelin's etiological-prognostic concept and created an extremely differentiated nosology of endogenous psychoses (Leonhard 1956, 1995). While retaining Kraepelin's prognostic dichotomy, he divided both groups (manic-depressive disease and dementia praecox) into various distinct illnesses. Kraepelin's manic-depressive grouping was subdivided into monopolar phasic psychoses, bipolar phasic psychoses (manic-depressive disease in a narrower sense) and cycloid psychoses. Dementia praecox was divided into the nosologically independent unsystematic and systematic schizophrenias according to Kleist-Leonhard (Fig. 1). Psychiatry today has access to completely independent classification systems of mental illness; atheoretical operationalized systems based on "expert consensus" on the one

Favorable long-term prognosis	Unfavorable long-term prognosis
Kraepelin: Manic-depressive disease	dementia praecox

Favorable long-term prognosis	unfavorable long-term prognosis
Leonhard: Affective cycloid psychoses psychoses	unsystematic systematic schizophrenia schizophrenia

Favorable long-term prognosis	unfavorable long-term prognosis
Bleuler: Manic-depressive group of schizophrenias disease	

Fig. 1. In his classification of endogenous psychoses, Leonhard retained Kraepelin's dichotomy of psychoses of long-term favorable and unfavorable prognoses. Bleuler's extension of the term "schizophrenia" – which also includes the cycloid psychoses according to Leonhard – makes prognostic diagnostics impossible

hand, and clinically-empirically founded nosology of the Wernicke-Kleist-Leonhard school on the other.

The nature – nurture controversy

The significance of heredity and environment with regard to the origin of mental illness was recognized from the very beginnings of psychiatry. One of the pioneers of psychiatry, Philippe Pinel (1745–1826), included heredity in addition to upbringing, irregularities in lifestyle and passions among the causes of mental illness (Ackerknecht 1985). Whether heredity or environmental factors would become more central to scientific interest was often determined by the given *zeitgeist*. Experience has shown, however, that the exclusive emphasis on only one of the two factors always leads to a dead end. Under the rubric "Wege und Irrwege der Genetik in der Psychiatrie" ("Right and Wrong Paths of Genetics in Psychiatry"), Propping once again demonstrated impressively how a too one-sided scientific viewpoint can lead to the depths of an ideologically motivated inhumanity (Propping 1989). Finally, the statement by Zerbin-Rüdin that it is not a question of "body or soul" and "nature or nurture" in the origin of mental disorders, but rather "body and soul" and "nature and nurture" (Zerbin-Rüdin 1974) is not a new development, but rather a return to the original position. It is pointless and "quite artificial and not biological" (Propping 1989) to attempt to determine the exact percentage attributable to heredity or the environment. In addition, the phenotype, or physical manifestation, is too far removed from the human genotype.

Outwardly visible human characteristics have usually developed
over a number of reactions and intermediate stages, from the genetic make-up to the final phenomenological state. The level of the
visible phenotype is preceded by the level of the gene products
(proteins), which is preceded by the chromosomal level and finally
the gene level (DNA level) (Propping 1984). In addition, some
hereditary characteristics are very resistant to environmental influences (such as height) while others are very susceptible (such as
body weight).

   On the other hand, to achieve an exact etiology and treatment
for mental illness it is absolutely necessary to determine the relative importance of heredity and/or environmental factors: whether,
for example, a hereditary disposition only occurs in certain environmental constellations or, on the other hand, if certain environmental constellations can possibly prevent the clinical manifestation of a hereditary disposition.

Methods of clinical genetics in psychiatry

Clinical genetics investigate genetic variability at the level of the
phenotype, i.e. distinct illnesses must first be clinically recognized
as such and described and defined by their pattern of symptoms.
Only when this step has been completed is it advisable to carry out
further research on the gene product, chromosome and/or gene.
Clinical genetics in psychiatry are based essentially on research with
families, adopted children and twins. The occurrence of certain illnesses within a family points to heredity, even though it does not
provide absolute proof. The disorder could possibly also be the
result of a pathogenic environment which has affected all members
of a family equally. On the other hand, the sporadic occurrence of
an illness in a family does not necessarily exclude heredity. A small
number or no siblings, or a rare recessive gene, for example, could
be the reason (Zerbin-Rüdin 1974).

   Adoption studies present a suitable means of separating the influence of the family environment and that of heredity with regard to
the occurrence of mental illness. There are essentially three strate-
gies which can be followed:

   1. Children of mentally ill parents are examined who have grown
up with healthy adoptive parents. Adopted children with healthy
biological parents are used as controls.

   2. The frequency of biological parents becoming ill is compared
with adoptive parents in the case of persons who were adopted at
an early stage and are now mentally ill.

3. Children of healthy biological parents are examined who have grown up with mentally ill adoptive parents. Adopted children of healthy biological parents whose adoptive parents are also healthy are used as controls.

An essential objection against the adoption method is that neither parents who give up their children for adoption nor parents who adopt children represent a group that is comparable to that of the normal population. As a result, the influence of genetic factors tends to be underestimated in adoption studies, since a higher rate of mental irregularities is expected even within the control groups (Propping 1989).

Twin research is the subject of this investigation, and will be explained in detail in the following chapters.

General remarks on twin origination

The introduction of the twin method in clinical genetics is most often traced back to Galton (1876). According to Vogel and Motulsky, however, Galton did not correctly recognize the essential aspect of the method – namely the existence of two different twin types. Dareste, however, had already reported on the distinction between monozygotic and dizygotic twins in 1874 to the Société d'Antropologie (Vogel and Motulsky 1986).

Biology of twin origination

Monozygotic and dizygotic twins

Dizygotic twins are conceived when two ova are ovulated in the woman's same cycle and are fertilized by two different sperms. Due to their common genetic background, dizygotic twins have about 50 per cent of their genes in common, like ordinary siblings. Each of the dizygotic twins always has his own chorion and his own amnionic cavity with surrounding membrane. Depending on the distance in which the two embryos are implanted in the uterus, the placenta can be separated or joined. If the placenta is joined, the separating membranes consist of amnion-chorion-chorion-amnion.

Monozygotic twins originate when a fertilized ovum in an early embryonic stage divides into two genetically identical daughter cells – a kind of asexual reproduction. This can occur up to about two weeks after fertilization. If the fertilized egg-cell up to the morula stage has already divided before the differentiation of the trophoblast (day five after fertilization), the twins will each have a separate chorion and amnion and the separating membrane will consist of amnion-chorion-chorion-amnion. These monozygotic twins have the same intrauterine conditions as dizygotic twins and cannot be distinguished from them at the embryonic stage. If

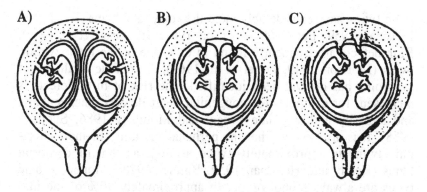

Fig. 2. (modified by Langman 1977). (A) Monozygotic twins with separate chorion and amnion. The division of the fertilized egg-cell has taken place before differentiation of the trophoblast (up to day five after fertilization). (B) Monozygotic twins with common chorion but separate amnion. The division of the inner embryonic cell mass takes place after the morula stage but before differentiation of the amnion (about day five – ten after fertilization). (C) Monozygotic twins with common chorion and amnion. Division takes place only after differentiation of the amnion (after day ten after fertilization)

the egg-cell divides after the morula stage, but before the differentiation of the amnions (between approximately days five and ten after fertilization), the twins will share a common chorion but have their own amnion. This separating membrane now consists of amnion-amnion. If division takes place only after the differentiation of the amnion (after day 10 after fertilization) the twins will have a common chorion and a common amnion, i.e. there is no separating membrane between the individuals (Fig.2). Twins with a separate chorion and amnion can therefore be dizygotic or monozygotic. Approximately one-third of monozygotic twins belong to this category. Twins with a common chorion, however, are always monozygotic. They generally have a separate amnion. Only a small number of monochorionic twins (ca. 4%) possess a common amnion (Bulmer 1970).

The effect of differences in intrauterine environment

An extremely important question for the subsequent discussion is whether monozygotic and dizygotic twins have different intrauterine environmental conditions. In a retrospective evaluation of ultrasound measurements carried out during the second trimester, Gilmore et al. (1996) found considerable differences in the brain development and body size of monozygotic twins. It has been known for a long time that twins have a significantly higher rate of birth defects than ordinary children (Benirschke and Kim 1973, Kohl and Casey 1975, Little and Bryan 1986), and that monozygotic twins have considerably more than dizygotic twins (Heady and Heasman 1959, Barr and Stevenson 1961, Stevenson et al. 1966). Birth defects are frequently found in only one of the monozygotic twins (Morison 1949, Fogel et al. 1965). A specific birth defect syndrome has not yet been identified. The risk of birth defects is increased, for example, for congenital heart defects, anencephaly, and cleft lip and/or palate (Vogel and Motulsky 1986, Spellacy 1988). It is also known that the perinatal mortality of monochorionic twins is approximately twice as high as that of dichorial twins (Heady and Heasman 1959, Bulmer 1970). Monochorionic twins are always monozygotic. In approximately 90% of the monochorionic placentas there are links in the blood circulation of the twins. In the case of arterial-venal shunts, which are not compensated for by other vascular links, one twin will bleed slowly into the other and will cause the so-called "twin transfusion syndrome". This occurs to some degree in 15–30% of monochorionic twins (Rausen et al. 1965, Campion and Tucker 1973). In the case of dichorial twins, placental vascular links between the two twins occur extremely rarely (Nicholas et al. 1957). The "transfusion syndrome" can lead to a chronic deficiency of oxygen and other important nutrients. One twin will be small and pale, with remarkably reduced levels of hemoglobin and serum proteins, and often suffers from hypoxia. The other twin is large and plethoric and often develops cardiac insufficiency with hydrops and polyhydramnios. Weight differences of up to 1000 grams can occur. The "transfusion syndrome" is generally viewed as the main cause of increased pre- and perinatal mortality and the increased rate of birth defects in monochorionic twins (Rausen et al. 1965, Campion and Tucker 1973, Vogel and Motulsky 1986, Spellacy 1988). In addition, in the case of monochorionic twins with only one amnion, the high prenatal mortality rate is also attributed to umbilical cord twisting (Bulmer 1970).

Frequency of twin births

According to Propping (1989), almost one in every 100 births in central Europe is a twin birth, making approximately every 50th person in the general population a twin. About 40% of twins are monozygotic, while the remaining 60% are dizygotic. Dizygotic twins are about equally the same or different sex. According to Weinberg (1902, 1909), the frequency of monozygotic and dizygotic twin rates is estimated according to the following formulas:

$$m = \frac{(G - U)}{N}$$

$$m = \frac{2U}{N}$$

m Monozygotic (identical) twin rate
d Dizygotic (fraternal) twin rate
G Number of same-sex twins
U Number of different-sex twins
N Number of all twin pregnancies

The formulas are based on the assumption that the sex ratio is 1:1, which is not quite correct, as a slightly higher number of boys are born than girls. Same-sex dizygotic twins also appear to be somewhat more frequent than different-sex twins, although the resulting difference is viewed as negligible (Bulmer 1970, Vogel and Motulsky 1986). The twin rates which can be calculated from these formulas are based entirely on the number of all twin pregnancies and not on the number of twin births. They therefore include both stillbirths and live births.

Frequency of monozygotic twins

Monozygotic twins occur at approximately the same frequency in all ethnic groups (approximately 3.5 for every 1000 births). Propping assumes that the rate of asexual division in early embryonic stages could be a universal constant (Propping 1989). According to Bulmer (1970) experiments with animals show that this phenomenon could arise as the result of a developmental inhibitor (e.g. lack of oxygen) in a very early embryonic stage. Nevertheless, the cause of the origin of monozygotic twins still remains unclear (LaBuda, et al. 1993).

Frequency of dizygotic twins

While the probability of having monozygotic twins is approximately the same for all women, the probability of having dizygotic twins is different for each woman. The relative frequency of dizygotic twins increases with the age of the mother, with a maximum between the ages of 35 and 39. After the age of 39, the frequency decreases (McArthur 1953, Bulmer 1970, Krüger and Propping 1976). One possible cause currently under discussion is the increased level of gonadotropin FSH in older women, which increases the frequency of multiple pregnancies. The frequent multiple pregnancies of women treated with hormones supports this hypothesis. The reduced twin rate in the last phase of the woman's reproductive years is attributed to the decline in polyovulation, despite a higher FSH level (Vogel and Motulsky 1986). Regardless of the woman's age the relative number of dizygotic twins also increases with the number of previous births (Bulmer 1970). A hereditary disposition is also assumed as dizygotic twin births occur more frequently in some families (Weinberg 1909, Bulmer 1970). This familial pattern is found exclusively on the maternal side. Male dizygotic twins or fathers of dizygotic twins do not have an increased probability of fathering twins. The differing frequency of dizygotic twins in different ethnic groups is also interpreted genetically. For example, the twin rate in the Caucasian population is approximately 8 per 1000 births, while the rate in the African population is almost twice as high, and the rate in the Asiatic population is barely half as much (Bulmer 1970, Propping and Krüger 1976, Vogel and Motulsky 1986).

Decline of dizygotic twin births in civilized countries

A decline in dizygotic twin births has been observed in all civilized countries since the mid 1950s. The current average younger age of the mothers is viewed only as a partial explanation. According to Propping and Krüger (1976), it seems more plausible to seek an explanation in the fact that a maternal disposition to increased fertility plays a much less important role than before in this era of birth control. Recent research in Italy and the USA has shown that the birth rate of dizygotic twins has stabilized again since the early 1980s (Allen and Parisi 1990).

The twins method in research

Characteristics of the twin constellation

Twins represent a distinctive social group (Schepank 1974, 1993). They are apparently less dependent on communication with the rest of their environment, since – if they grow up together – they always have a partner of the same age. The attachment to each another is much more intense with monozygotic twins than with dizygotic twins. While monozygotic twins often strive to attain similarity and identification with each other, dizygotic twins tend to show more competitive behavior and a wish to distinguish themselves from their partner (Bischoff 1959, Bracken 1969, Schepank 1993). Generally, the identification with the partner appears to be stronger in female monozygotic twins than in male pairs (Vogel and Motulsky 1986). However, there are a significant number of dizygotic twins who also exhibit a strong degree of attachment and a pronounced aspiration for similarity. Conversely, it is possible for monozygotic twins to be less attached to each other and show a strong wish to distinguish themselves from each other (Schmidt 1986). It is even possible for an extreme animosity to develop between the twins. Attempts to attain similarity and/or difference are also influenced by the environment. In the past, it was more often the case that the identical nature of monozygotic twins was additionally emphasized by clothing, etc., while currently many educators place more value on emphasizing their individuality (Friedrich and Kabat vel Job 1986).

Specific role assignments often occur with twins. One twin is often the speaker and/or the more dominant of the two, while the other twin is more subordinate and less independent. Shields (1962) examined 44 pairs of monozygotic twins who grew up separated and compared them to a control group of 44 pairs of monozygotic twins who grew up together. He discovered that the dominant/subordinate axis was the most distinctive characteristic of monozygotic twins. Interestingly, Shields found a significant relationship between a higher birth weight and the leadership role in the pair (Shields 1962).

Von Verschuer was able to show that the so-called "twin fate", which means that an illness or death of a monozygotic twin automatically results in the same fate for the other, occurs only very rarely. "Individual destiny remains unpredictable" and "the constant photographic similarity of twin fates is a myth" (Zerbin-Rüdin 1974).

The methodological basis of twin research

Classical twin research

Classical twin research is based methodologically on the comparison of monozygotic and dizygotic twins and assumes the same environmental factors for both twin types. According to Galton's law, heredity is indicated when the correspondence of a certain trait is greater between monozygotic twins than between dizygotic twins. In the case of non-heredity, monozygotic and dizygotic pairs do not differ considerably in their correspondence (concordance) or lack of correspondence (discordance) for the trait. All phenomenological differences between monozygotic twins are attributed to the influence of the environment (Zerbin-Rüdin 1980). Galton's law should, however, be applied with certain restrictions. In the case of monozygotic twins, unfavorable prenatal conditions could lead to increased concordance as well as increased discordance for hereditary and non-hereditary illnesses (see pages 7, 8). Therefore, birth weight and complications during pregnancy and birth should always be taken into consideration in twin studies (Campion and Tucker 1973). Also the fact that some monozygotic twins are concordant for an illness while others are discordant can be accounted for by various explanations: 1. the manifestation of a hereditary illness is influenced considerably by non-genetic factors, or 2. there is both a hereditary form as well as a non-hereditary form of the illness. In order to be able to make this distinction it is necessary to compare the empirical familial pattern of an illness of concordant and discordant pairs. If there is no genetic type of an illness, the morbidity risk of close relatives of discordant pairs is not greater than in the normal population. If non-genetic factors must be present in order for the illness to become manifest, the morbidity risk of relatives in both concordant and discordant twins is approximately equal (Vogel and Motulsky 1986).

A repeated problem of twin studies, namely that higher rates of concordance between monozygotic twins in comparison with dizygotic twins is primarily attributed to the special twin constellation (see page 11), has been researched specifically by several authors. The results of some of these studies are summarized in Table 1.

The studies essentially prove that the basic assumption of twin research – greater concordance rates in monozygotic as compared with dizygotic twins, based on predominately genetic influence – continues to be valid according to current scientific findings. A second objection against the twins method is that the absolute genetic identity of monozygotic twins is questionable. There is noth-

Table 1. Studies of possible genetic interpretations of the twin method (modified by Kendler 1983 and Propping 1989)

Author	Method and results
Scarr (1968):	The similarity of different variables was investigated by comparing the actual with the supposed (parent's opinion) zygosity (52 MZ pairs): *MZ which were believed to be DZ were viewed as much more similar in almost all variables than DZ which were believed to be MZ.*
Munsinger and Douglass (1976):	The actual zygosity was compared to the believed zygosity with respect to language ability (74 pairs): *the degree of similarity corresponded almost exactly to the actual and not the believed zygosity.*
Plomin et al. (1976):	The relationship between physical likeness and similarity in personality was investigated in same-sex twins (288 pairs): *no correlation between the variables was found in the MZ pairs.*
Matheny et al. (1976):	The investigation concerned the relationship between physical likeness and similarity in various achievement tests in same-sex twins (191 pairs): *no correlation was found in the MZ pairs.*
Loehlin and Nichols (1976):	The investigation concerned the relationship between the similarity in environment in childhood and personality as well as certain abilities (850 pairs): *no correlation was found.*
Vandenberg and Wilson (1979):	The investigation concerned the relationship between the degree of similarity of twins using various achievement tests (45 MZ, 37 DZ): *no correlation was found.*
Matheny (1979):	The investigation concerned the similarity in the Stanford-Binet IQ Test in comparison with believed zygosity (172 pairs): *the similarity in IQ reflected only the actual and not the believed zygosity.*
Scarr and Carter-Saltzman (1979):	The relationship between achievement and personality test results was investigated by comparing the actual and the believed (opinion of the twins) zygosity (400 pairs): *the similarity in the test results reflected almost exclusively the actual zygosity.*
Kendler and Robinette (1982):	The investigation concerned the correlation between the physical likeness and concordance for schizophrenia (164 MZ pairs in which at least one was schizophrenic): *no correlation was found.*

ing to indicate genetically dissimilar division in humans, which can occur in primitive forms of life. Human monozygotic twins correspond in all hereditary traits investigated (Bulmer 1970, Zerbin-Rüdin 1980, Vogel and Motulsky 1986).

Methodological variants

Interesting methodological variants in twin research include:

The comparison of partners in monozygotic pairs. In the case of monozygotic partners it is possible to control genetic as well as environmental factors, where one twin always functions as the control person.

The comparison of twins with other siblings. The comparison of dizygotic twins with other siblings makes it possible to investigate the social factors of the twin constellation, since dizygotic twins and ordinary siblings have the same genetic conditions, e.g. 50% of the same genetic material.

The comparison of monozygotic twins who have grown up together and apart. This is often viewed as the "cream of the crop" of twin research because it investigates genetically identical persons having grown up in different environments. However, these situations are very rare. Concerning schizophrenia, for example, Propping (1989) calculated a probability of 1 : 6,250,000 that an individual belongs to a monozygotic twin pair, is diagnosed with schizophrenia and is separated from his twin in early childhood.

Definition of terminology

Index-twin and co-twin. An index-twin is the twin that was hospitalized for psychiatric disease, ascertained during a systematic review of medical files. If both partners of a twin pair were hospitalized, both partners will be index-twins. If only one twin was hospitalized for psychiatric disease, his partner is designated as the "co-twin", even if he is also ill.

Zygosity (zygosity diagnosis). Identical twins are generally designated as monozygotic (MZ) and fraternal twins as dizygotic (DZ). A lack of valid methods to determine zygosity inhibited twin research for quite some time, until Siemens (1924) successfully developed a reliable method of "polysymptomatic similarity comparison". Currently, however, more importance is attached to the determination of genetic polymorphisms, such as blood group characteristics (Kabat vel Job 1986) or DNA polymorphisms (Erdmann et al. 1993). Particularly large twin studies still occasionally use only postal questionnaires, as many studies have shown that the results of this questionnaire method correspond to approximately 95% of those from the blood-serological method (Cerderlöf et al. 1961, Nichols 1965, Torgersen 1979, Sarna and Kaprio 1980).

Concordance and discordance. When both partners of a twin pair exhibit a certain trait (such as a specific illness), this pair is considered to be concordant (C) for this trait (the illness). If this trait (the illness) is found in only one partner, the pair is discordant (D) for the trait (the illness). Various methods are applied in the concordance calculation, each of which can produce different values. Distinctions can be made based on *pair-wise, proband-wise and case-wise* concordance rates (Fig. 3). In the *pair-wise calculation* the percentage of the concordant pairs is simply calculated from the total number of investigated pairs. The *proband-wise* method is based on the number of index-twins during the twin survey. If both partners of a pair are index-twins, the pair is counted twice for the calculation. If the index-twins are ascertained independently of one another in a systematic study, the *proband*-based method is the best choice. In this case each ascertained person is included in the calculation, and the calculated concordance rate can be compared directly with the empirical repetition figures taken from the investigations of families (Allen et al. 1967, Slater and Cowie 1971, Allen and Hrubec 1979, Gottesman and Shields 1982, Propping 1989, McGue 1992). The *pair-wise* method provides the advantage of simple calculation and permits the direct application of statistical analyses (e.g. chi^2-Test). If only one partner of a pair is an index-twin in all cases within a study, the *pair-wise* and *proband*-wise calculations result in identical concordance rates. If in some cases both partners of a pair are index-twins, however, the *proband*-wise concordance rate will always be higher than the *pair-wise* rate.

With the *case-wise calculation*, all ill individuals who have an ill partner are counted. This then does not depend on whether or not they are index-twins. In this way all concordant pairs are automatically counted twice. The *proband*-wise and the *case-wise* calculations show the same result when all individuals in the concordant pairs are index-twins.

1. Pair-wise concordance: 12 of the 26 pairs are concordant (C) and 14 are discordant (D). The concordance rates are calculated on the basis of the following formula:

$$\frac{C}{C+D} = \frac{12}{12+14} = \frac{12}{26} = 46.2\%$$

2. Proband-wise concordance: Of the 52 individuals coming from 26 twin pairs, 31 of these were identified as index-twins independently of one another. In five pairs both partners were index-twins. Both partners in these five pairs were counted in the investigation, i.e. there were only 12 concordant pairs, but 17 index-twins who had an ill partner: Zw(C). 14 index-twins had a healthy partner: Zw(D).

The concordance rate is calculated using the following formula:

$$\frac{Zw}{Zw(C)+Zw(D)} = \frac{17}{17+14} = \frac{17}{31} = 54.8\%$$

3. Case-wise concordance: Irrespective of the survey, there were 24 ill individuals with an ill partner among the 52 individuals (26 pairs), i.e. in every concordant pair there were two individuals with an ill partner (2C). Fourteen ill individuals had a healthy partner (D). The concordance rate is calculated from the following formula:

$$\frac{2C}{2C+D} = \frac{24}{24+14} = \frac{24}{38} = 63.2\%$$

Concordant pairs   **Discordant pairs**
(C)   **(D)**

A Index-twin
B Ill co-twin
C Healthy co-twin

Concordant		Discordant	
A ———— A		A ———— C	
A ———— A		A ———— C	
A ———— A		A ———— C	
A ———— A		A ———— C	
A ———— A		A ———— C	
A ———— B		A ———— C	
A ———— B		A ———— C	
A ———— B		A ———— C	
A ———— B		A ———— C	
A ———— B		A ———— C	
A ———— B		A ———— C	
A ———— B		A ———— C	
		A ———— C	
		A ———— C	

Fig. 3. (modified by Propping 1983): The figure shows 26 twin pairs. Twelve pairs (left half of Fig. 3) are concordant (C), i.e. correspond with respect to the trait, and 14 pairs (right half of Fig. 3) are discordant (D), i.e. do not correspond with respect to the trait. In this example a systematic twin survey resulted in 31 index-twins (A). As the figure shows, there are seven ill twins who were not ascertained in the survey, i.e. who are ill co-twins (B). Fourteen co-twins are healthy (C)

Limited representative twin survey. This method of twin survey is applied most often. All twins from a population of persons with the trait (e.g. archive of stationary psychiatric patients) are systematically screened. The survey is successful when the frequency of twins in this population (e.g. schizophrenic subjects) corresponds approximately to the frequency of twins in the normal population.

Unlimited representative twin survey. With this method every twin in the general population is screened and examined with respect to one trait (e.g. schizophrenia). The entire register of births for a region over a period of several years must be checked. This method can only be used when a twin register and a register of the population with the investigated trait (e.g. psychotic disorders) are kept in addition to a birth register. In this case it is possible to compare the twin register with the register for the specific disorder. This was practiced for psychiatric disorders over a period of time in Denmark, Norway and Finland.

Hereditariness. Hereditariness is the designation of the estimated genetic proportion of the total observed variance of an evenly distributed trait (Propping 1989). It is worthwhile to calculate the hereditariness index of all quantitative traits of unknown origin (Allen 1979). The indication on this index is useful when a multiple-factor heredity in the case of schizophrenia is assumed. The model of multiple-factor heredity assumes polygenic heredity, with or without major genetic effects, and with or without a threshold value. According to Holzinger (1929), the hereditariness index can be calculated using the following formula:

$$H = \frac{\text{concordance rate MZ} - \text{concordance rate DZ}}{1 - \text{concordance rate DZ}}$$

With a threshold model, the following formula, according to Allen (1979), describes the genetically determined factor more clearly:

$$H = 1 - \frac{\text{concordance rate DZ}}{\text{concordance rate MZ}}$$

The hereditariness index should not be overestimated. Especially in twin studies with a small number of cases, both the genetic and the non-genetic percentages can be overemphasized. Hereditariness estimates do not provide insight into the extent of pair screening, dominance effects and heredity-environment interaction (Vogel and Motulsky 1986, Kendler 1989). The model of multiple-factor inheritance is often viewed as a stopgap (Propping 1989). The number of genes assumed to be involved can be mathematically increased or decreased until the model more or less applies (Zerbin-Rüdin 1974). This means that it is possible to explain everything with this model without making a truly definitive statement.

Age adjustment. Propping (1989) advises adjusting for age only to an extent in twin studies. When monozygotic pairs are concordant for schizophrenia, this is generally true in most cases within a period of four years (Kallmann 1946, Hoffer and Pollin 1970, Belmaker et al. 1974). In this study, however, schizophrenic psychoses are differentiated and divided into different subgroups so that age adjustment must be taken into account for the purpose of completeness. The classical method according to Strömgren (1936) was used for the purposes of calculation, since this method has been applied in most of the twin studies conducted so far. The method is based on the positive correlation between the ages of onset of schizophrenia in siblings as determined statistically in a large test group. Every test person who was healthy at the time of the investigation is accounted for with a reference rate between 0.00 (not counted) and 1.00 (fully counted), based on sex, age and the age of his sibling at the onset of the illness (Strömgren 1936).

Classical twin studies in schizophrenia research

The important twin studies in schizophrenia research up to now are summarized in Table 2. The pair-wise calculation forms the basis of the concordance rates in this table (see pages 15, 16). Table 3 shows the proband-wise concordance rates for the studies which supplied a sufficient amount of information. In particular, the concordance rates of the monozygotic twin pairs show considerable fluctuation in both tables. Various diagnostic interpretations are particularly to blame for these fluctuations (Shields and Gottesman 1972), in addition to sources of error, such as the methods of the twin survey, determination of the zygosity, and the statistical treatment of data (Kringlen 1971). The diagnostic concept affects the results of the concordance rate in two ways: In the assignment of the index case to a specific diagnostic category on the one hand, and in the diagnostic assessment of the co-twin on the other. Despite these differing concordance rates, the calculation of hereditariness according to Allen (1979) (see page 17) produces consistent values of frequently more than 0.70 in almost every study (Table 3).

Gottesman and Shields (1972) prepared summaries of the medical histories of their 114 twin subjects. These medical histories

Table 2. Twin studies in schizophrenia research: Pair-wise concordance rates for diagnosed and probable schizophrenia (the concordant pairs/total number of pairs are shown in parentheses)

Author	Monozygotic pairs	Dizygotic pairs
Luxenburger		
(1928)	71.4% (10/14)	0% (0/33)
(1930)	66.7% (14/21)	0% (0/37)
(1934)	33.3% (9/27)	
Rosanoff et al.		
(1934)	61.0% (25/41)	10.0% (10/101)
Essen-Möller		
(1941)	63.6% (7[1]/11)	14.8% (4/27)
(1970)	72.7% (8[1]/11)	
Kallmann		
(1946)	68.9% (120/174)	14.7% (76/517)
	85.8%[2]	
	69.0%[3]	
Slater		
(1953)	64.8% (24/37)	8.9% (10/112)
	76.0%[2]	
Kallmann and Roth		
(1956)	70.6%[4] (12/17)	17.1% (6/35)
	88.2%[5] (15/17)	22.9% (8/35)
Inouye		
(1961)	60.0% (33/55)	11.8% (2/17)
Tienari		
(1963)	0 (0/16)	4.8% (1/21)
(1968)	35.7% (5/14)	
Gottesman and Shields		
(1966)	54.2% (13/24)	9.1% (3/33)
(1972)[6]	50.0% (11/22)	9.1% (3/33)
Kringlen		
(1967)	38.0% (21/55)	13.8% (13/94)
Fischer et al.		
(1969)	47.6% (10/21)	19.5% (8/41)
Pollin et al.		
(1969)	13.8% (11/80)	4.1% (6/146)
Hoffer and Pollin		
(1970)	40.0% (32/80)	11.0% (16/146)

Table 2. (continued)

Kendler and Robinette (1982)	18.3% (30/164)	3.4% (9/268)
Onstad et al. (1991)	33.3%[7] (8/24)	3.6% (1/28)

[1] One of the co-twins committed suicide.
[2] Age adjusted.
[3] Corrected according to Shields et al. 1967.
[4] "Preadolescent schizophrenia": co-twin also developed schizophrenia before age 15.
[5] "Preadolescent schizophrenia": co-twin developed schizophrenia before and after age 15.
[6] Consensus diagnoses of 6 diagnosticians, based on medical histories.
[7] DSM-III-R criteria.

Table 3. Proband-wise concordance rates and hereditariness index according to Allen (1979), in studies which permitted these calculations

Author	Monozygotic pairs	Dizygotic pairs	Index for hereditariness
Luxenburger (1928)	76.5%		
Essen-Möller (1941)	63.6%	14.8%	0.77
Slater (1953)	78.0%	23.0%	0.71
Inouye (1961)	60.0%	18.2%	0.70
Kringlen (1967)	44.9%	14.6%	0.68
Fischer et al. (1969)	60.9%	27.9%	0.54
Gottesman and Shields (1972)	57.7%	11.8%	0.80
Kendler and Robinette (1982)	30.9%	6.5%	0.80
Onstad et al. (1991)	48.4%[1] 67.7%[2]	3.6% 28.6%	0.93 0.56

[1] Schizophrenia diagnosis only.
[2] Included are diagnoses of the entire "schizophrenic spectrum".

Table 4. Effects of various schizophrenia criteria on the concordance rates of the twin study by Gottesman and Shields (Shields and Gottesman 1972)

	Monozygotic pairs	Dizygotic pairs	MZ/DZ
Gottesman and Shields	54.2% (13/24)	9.1% (3/33)	6.0
Consensus of six psychiatrists	50.0% (11/22)	9.1% (3/33)	5.5
Very broad criteria	58.3% (14/24)	24.2% (8/33)	2.4
Very strict criteria	20.0% (3/15)	13.6% (3/22)	1.5

were then diagnosed by six clinicians from three different countries (England, USA, Japan) without knowledge of the zygosity. Possible diagnoses were: schizophrenia; schizophrenia?; other diagnoses; normal. The frequency of the diagnosis of schizophrenia fluctuated between 43 (England) and 77 (USA), and within England alone between 43 and 64. Consensus diagnoses were then compiled. Two monozygotic pairs were excluded since they did not include schizophrenic index patients. Ultimately, the pair-wise concordance rate for "schizophrenia and schizophrenia?" amounted to 50% for monozygotic twins, and 9% for dizygotic twins. The most telling difference in the concordance rates between monozygotic and dizygotic twins was achieved with "middle-of-the-road" criteria for schizophrenia, which according to Shields and Gottesman (1972) are seen as rather broad in Europe, but which are seen in the USA as being rather strict. The difference in the concordance rates between monozygotic and dizygotic pairs (ratio monozygotic/dizygotic) was considerably smaller both in very broad as well as in very strict schizophrenia criteria (Table 4).

Severity of illness and concordance rates

In several twin studies evidence is found on distinctly different concordance rates with differing severity of the illness among the index twins (Table 5).

In a study by Kallmann (1946), schizophrenic monozygotic index-twins with a moderate to severe mental disorder had in every case (100%) an ill co-twin. Index-twins having a mild course of schizophrenia and no mental impairment had an ill co-twin in only 26% of the cases. Of the dizygotic pairs, 17% of the severely

Table 5. Concordance rates of monozygotic twin pairs in relation to the severity of the disorder in the index-twin

Author	Index-twin moderately or severely ill	Index-twin mildly or temporarily ill
	concordant to	concordant to
Kallmann (1946)	100%	26%
Inouye (1961)	77%	39%
Gottesman and Shields (1966)	77%	27%
(1982)	75%	17%
	91%[1]	33%[2]

[1] Application of Slater's criteria "nuclear".
[2] Application of Slater's criteria "nonnuclear".

ill index-twins had an ill co-twin, but only 2% of the mildly or temporarily ill index-twins had an ill co-twin.

In the study by Inouye (1961), the 23 index-twins with "chronically progressive schizophrenia" had 17 ill partners (74%), and seven index-twins with a "remittent, constantly reemerging schizophrenia" had six ill partners (86%). Taken together, the concordance rate of these prognostically unfavorable schizophrenic psychoses was 77% (23 of 30 pairs were concordant). The concordance rate in the group of index-twins with a "chronically mild or temporary schizophrenia" was considerably less, namely only 39% (nine of 23 pairs were concordant). Gottesman and Shields (1966, 1982) also investigated the relationship of the severity of the illness in the index-twin and the risk of it developing in the co-twin. They selected the length of hospitalization and the level of impairment (e.g. ability to work) as indicators for the severity of the disorder in the index-twin. The co-twins of severely ill index-twins had a three-times greater likelihood of developing the disorder, compared with the "mildly" ill. The application of Slater's sub-typing in "nuclear versus nonnuclear test persons" leads to the concordance rates of 91% (index-twin "nuclear schizophrenic"), compared to 33% (index-twin "nonnuclear schizophrenic").

Concordance/discordance and familial pattern

In an analysis of Slater's twin series, Rosenthal (1959) determined that there was a marked tendency for additional schizophrenic psychoses to occur in families of monozygotic twins which were concordant for schizophrenia. In four of five families for which a sufficient amount of data was available, other severely schizoid and/or schizophrenic members were present also. The schizophrenia ran a prognostically unfavorable long-term course with these concordant pairs. In contrast, monozygotic pairs which were discordant for schizophrenia had in no cases other family members who were ill, even though the total number of relatives was larger. Furthermore, the schizophrenia generally ran a favorable course in these cases. Gottesman and Shields (1966) did not find additional cases of schizophrenia in the families of concordant or discordant pairs. Kringlen (1967) reported that the discordant pairs showed a slight tendency towards a lower number of ill family members. Also Lewis et al. (1987) reported that the discordant pairs tended to have fewer psychotic relatives of the first and second degree (in eight of 18 pairs) than the concordant pairs (in nine of 13 pairs) in a group of 31 monozygotic twin pairs. On the other hand, Onstad et al. (1992) found no difference in the familial pattern when comparing 16 discordant with eight concordant monozygotic twin pairs.

Concordance/discordance and handedness

Boklage (1977) examined the handedness of monozygotic and dizygotic twin pairs, where at least one twin partner had been diagnosed with schizophrenia. In the group of monozygotic twins, there were three times as many "non-right-handed" partners as in the group of dizygotic twins. Boklage found a highly significant concentration of "non-right-handed" partners among schizophrenic twins of monozygotic pairs compared with healthy twins of monozygotic pairs. The "non-right-handed" twins were especially predominant in monozygotic pairs which were discordant for schizophrenia. Of 12 monozygotic twin pairs with two right-handed partners, 11 pairs (92%) were concordant for the diagnosis of schizophrenia. On the other hand, of 13 pairs which were discordant for schizophrenia, there were no fewer than 12 pairs which had at least one "non-right-handed" partner. Furthermore, schizophrenic twins from pairs with at least one "non-right-handed" partner showed, without exception, a considerably more favorable course of illness compared with schizophrenics from pairs with two right-handed partners. Luchins et al. (1980) also found that left-handedness was not associated with nuclear schizophrenia, but rather with a more favorable form of the disorder (nonnuclear variant).

Lewis et al. (1989) were not able to reproduce Boklage's results, although the authors adhered strictly to his methods. Torrey et al. (1993a) were also unable to confirm Boklage's findings.

Concordance rate of dizygotic twins and other sibling morbidity rate

Since dizygotic twins cannot be distinguished from ordinary siblings in their genetic make-up, it would be genetically expected that the likelihood of their developing an illness was also identical. However, virtually consistent findings in systematic twin studies, which also investigate the morbidity rate of other siblings, show that the concordance rates of dizygotic twins are higher than the morbidity rates of other siblings (Table 6). Kringlen (1990) attributes this to the twin constellation, without being conclusive.

Monozygotic twins reared apart

No systematic study has been carried out of monozygotic twins with a schizophrenic index-twin who were reared apart (see page 14). Gottesmann and Shields (1982) report on 14 pairs which were surveyed within various systematic studies and who were known to have been reared apart. Nine of these pairs were concordant for schizophrenia (pair-wise concordance 64%). This result can be easily integrated into the concordance rates of systematic twin studies. Table 7 shows a list of monozygotic twin pairs from systematic studies who were separated before age five. Here as well, 67% of the pairs were concordant for schizophrenia. Single case studies must be regarded with caution, since their selection and

Table 6. Concordance rates of dizygotic twin pairs compared with the morbidity rates of schizophrenia (in percentage) in other siblings (according to Kringlen 1990)

Author	Concordance rate of twins (DZ)	Morbidity rate of other siblings
Luxenburger	14.0%	11.8%
Kallmann	10.3%	10.2%
Slater, Shields	11.3%	4.6%
Kringlen	8.1%	3.0%
Fischer	26.6%	10.0%

Table 7. Concordance/discordance of monozygotic twins already separated before age 5 (according to Kringlen 1990)

Author	Age at separation	Concordant	Discordant
Slater and Shields	birth	1	–
Tienari	3 years	–	1
Kringlen	2 years	–	1
	3 months	1	–
Inouye	7 days	1	–
	3 years	1	–

publication depend greatly on the subjective interest of the author (Propping 1989).

Frequency of schizophrenia in twins

A number of authors have investigated whether schizophrenia occurs more frequently in twins than in the general population. In systematic twin studies the frequency of twins in the total population of schizophrenic subjects was not greater than the frequency of twins in the general population (Gottesman and Shields 1982). Schizophrenic psychoses in monozygotic twins were also no more frequent than in dizygotic twins (Luxenburger 1928, Essen-Möller 1941, Harvald and Hauge 1965, Kringlen 1967). Allen and Pollin (1970) (*Table 8*) and Kendler et al. (1996) were able to confirm this in representative studies.

Table 8. Frequency of schizophrenia in twins compared with the general population (according to Allen and Pollin 1970)

	Frequency of schizophrenia in %
General population	1.14
Monozygotic twins	0.97
Dizygotic twins	1.22
Zygosity unknown	1.22

Studies of monozygotic twin pairs discordant for schizophrenia

Prenatal developmental disorders

In the fourth and fifth months of fetal development, ectodermal cells begin to migrate towards the upper limbs in order to form, among other things, the skin of the hands (Hamilton et al. 1972, Schaumann and Alter 1976). This migration of skin cells is genetically programmed (Holt 1968) and is extremely sensitive to non-genetic stressors (Wakita et al. 1988, Newell-Morris et al. 1989). Healthy monozygotic twin pairs have almost completely identical fingerprints, and only different intrauterine developmental conditions can bring about modifications (Hamilton et al. 1972, Schaumann and Alter 1976). It is known that a reduced amount of ridges on the skin surface of the fingers can be influenced by intrauterine anemia, anoxia, ischemia, maternal alcohol or drug abuse or other exposure to toxins, as well as by the transfusion syndrome. An increased amount of ridges can be the result of a fetal edema caused by, for example, an infection (Achs et al. 1966, Alter and Schulenberg 1966). Exposure in utero to infections and other damaging toxic conditions will not necessarily affect both twins (Goedert et al. 1991, Davis et al. 1995). In 24 schizophrenic index-twins of discordant monozygotic twin pairs, Bracha et al. (1991) found significantly more minute malformations of the hands compared with the healthy co-twins, which could be traced to developmental disorders in the fourth/fifth month of pregnancy. The authors compared the fingerprints of 23 monozygotic twin pairs who were discordant for schizophrenia. They found that in approximately one-third of the discordant pairs, the intra-pair difference in the amount of ridges on the skin surface of the fingers was significantly greater than in the healthy control twin pairs (Bracha et al. 1992). In a later investigation the research group reported developmental disorders in the thirteenth to fifteenth week of gestation in the schizophrenic twins but not in the healthy partners (Davis and Bracha 1996). This was evaluated as an indication that prenatal affective, environmental, and exogenous stressors could play an important etiological role in at least a part of the psychoses that belong to the term "schizophrenia". Van Os et al. (1997) came to the same conclusions, although they were not able to reproduce unequivocally the above-mentioned results.

Peri- and postnatal findings

In a meta-analysis of six twin studies (Slater 1953, Inouye 1961, Tienari 1963, Kringlen 1967, Fischer et al. 1969, Gottesman and

Shields 1972), Gottesman and Shields (1976) reported that the schizophrenic or severely ill twin had the lower or the higher birth weight just as often as the healthy or mildly ill twin. Reveley et al. (1984) and Onstad et al. (1992) also found no significant difference in birth weight, order of birth, or the physical condition after birth between the schizophrenic index-twin and the healthy co-twin. Lewis et al. (1987) reported that in 13 discordant pairs, the ill twin weighed less at birth in seven cases, weighed more in three cases, and had the same weight in three cases compared with the healthy partner (not significant). In this study, however, the average weight difference was considerably greater in the discordant pairs (312 ± 331 g) than in the concordant pairs (78 ± 33 g) (p = 0.07).

The research group consisting of Pollin, Stabenau, Mosher and Tupin exclusively selected distinctly discordant pairs. Their findings can therefore be compared only with some reservations to other studies which are based primarily on systematic twin surveys. In 1965 this research group found that in five discordant pairs, the schizophrenic twin always weighed less and was the weaker twin at birth, to the extent that the mothers had been very concerned about the child's survival. This twin's early childhood and childhood development were clearly delayed compared with the non-schizophrenic partner (Pollin et al. 1965). The authors reported in 1966 that in 11 discordant pairs the schizophrenic twin was the lighter twin at birth in 11 cases (between 15 g and 793 g) and was the second born in eight cases (Pollin et al. 1966). Stabenau and Pollin (1967) found in earlier twin studies of discordant pairs that the ill partner was often smaller, weaker and lighter at birth and was often affected by asphyxia or other birth complications. In 1971 Mosher et al. reported that in 15 discordant pairs, the ill index-twin showed neurological irregularities significantly more often, had more difficulties during birth, and was the lighter twin in 12 cases. The case reports from Pollin's research group suggest that these index-twins suffered primarily from remittent psychoses with only mild or no residual symptoms.

Neuroradiological findings

In a CT study of seven monozygotic twin pairs which were discordant for schizophrenia, Reveley et al. (1982) found a considerable intra-pair difference in ventricle width. The schizophrenic subjects had significantly wider ventricles than their healthy partners and the healthy monozygotic control pairs. The monozygotic control pairs (n = 11) also exhibited a high rate of correspondence in ventricle width in contrast to the dizygotic control pairs (n = 8). This suggests that there is a genetic determinant in ventricle width. The primarily genetically determined size of the human brain was recently reconfirmed by Bartley et al. (1997). In an investigation with monozygotic pairs which were discordant (n = 12) and concordant (n = 9) for schizophrenia, as well as healthy monozygotic control pairs (n = 18), the Reveley group also reported that ventricle dilation was limited to schizophrenic index-twins with a negative family history for "major psychiatric disorders" (not restricted to schizophrenia). Seven index-twins with a positive family history had significantly smaller total ventricle volumes than the 14 index-twins with a negative family history. Of six schizophrenic index-twins with additional birth complications, all had larger ventricles and a negative family history, and in five of these cases the co-twin was healthy (Reveley et al. 1983, 1984). In a variance analysis, it was then shown that the discordance/concordance factor had no significant effect on the ventricle width. The only significant predictor for ventricle size was the family's medical history. At this point it would also be appropriate to mention the neuroradiological findings on the famous "Genain quadruplets", who are monozygotic and concordant for schizophrenia. The father of the quadruplets was described as a severe psychopath (Rosenthal 1963). Interestingly, all of the quadruplets showed narrower ventricles than normal control persons in the examination by Buchsbaum et al. (1984).

Casanova et al. (1990) carried out an MRI study of the corpus callosum with 12 discordant monozygotic twins. In their comparison of ill and healthy twins they found no morphological irregularities, but did find differences in the form of the front and middle segments of the Corpus callosum, which also resulted from hydrocephalic ventricles.

Suddath et al. (1990) examined 15 discordant pairs using MRI. In 12 of the 15 pairs it was possible for a neuroradiologist, having no information on psychiatric diagnoses, to identify the schizophrenic index-twin solely on visual inspection of the MRI images, on the basis of additional cerebrospinal fluid. In two cases no difference in the MRI images were recognized in this manner, and in one case the healthy partner was named. Quantitative analyses showed (here compared with a mentally healthy co-twin) that the left hippocampus in 14 of 15 index-twins and the right hippocampus in 13 index-twins were significantly smaller at the level of the pes hippocampi. The lateral ventricles were significantly wider on the left side in 14 index-twins, and on the right side in 13 index-twins. The third ventricle was also significantly wider in 13 index-twins. Similar results could not be found in the seven monozygotic control pairs.

Weinberger et al. (1992a) compared male monozygotic discordant (n = 8) pairs with concordant (n = 7) pairs. No differences in the average width of the fourth ventricle were found. When comparing the twins with each other, the schizophrenic index-twins had significantly wider third ventricles, while the average ventricle width of the concordant twins did not differ. The concordant twins also had significantly wider third ventricles than the healthy twins of the discordant pairs. That the width of the ventricles does not correspond with the concordance/discordance factor has also been confirmed by the findings of Reveley et al. (1983).

In another group of monozygotic pairs who were discordant for schizophrenia (n = 7), Weinberger et al. (1992a) used MRI to compare the intrapair-similarity of the cortical gyral formation with a control group of mentally healthy monozygotic twin pairs (n = 5). They found that the similarity of the gyral pattern in the healthy control group (n = 5) was more pronounced than in the pairs with a schizophrenic twin. According to a study by Bartley et al. (1997) the cortical gyral patterns are decisively influenced by genetic factors, although their final developmental stage depends primarily on non-genetic factors.

The Weinberger group also reported that the greater the difference in the ill twin partner from the healthy co-twin in the volume of the left hippocampus (unfortunately the report did not include information on the phase of the disorder of the test persons: acute, remittent, residual, etc.), the more limited was the physiological pre-frontal activation (measurements based on regional cerebral blood flow rCBF) during the Wisconsin Card Sorting Test (Weinberger et al. 1992b).

Biochemical and other findings

Biochemical findings. Murphy and Wyatt (1972) and Wyatt et al. (1973a) reported that the activity of the mono-amino oxidase in blood platelets of monozygotic schizophrenic twins was lower than in control cases. The non-schizophrenic co-twins also showed this reduced enzyme activity.

Reveley et al. (1983b) compared the activity of the mono-amino oxidase of four groups, matched for age and sex: 10 monozygotic twin pairs which were discordant for schizophrenia, 20 healthy monozygotic twin pairs, 20 healthy dizygotic twin pairs and 20 unrelated control persons. In the monozygotic twin groups a significant ($p < 0.01$) intrapair-correlation of the MAO activity was found. Discordance for schizophrenia as well as medication (neuroleptics in the case of the schizophrenic index-twin vs. no medication in the case of the healthy co-twin) had no essential influence on the activity of the MAO. There was no intrapair-correlation with the dizygotic twins. The activity of the MAO was significantly reduced in the group of monozygotic twins which were discordant for schizophrenia (i.e. both the ill and the healthy twin), both compared with the entire group of control persons ($n = 60$) as well as with every individual group of control persons. Since no intrapair-differences occurred in the monozygotic discordant pairs, the reduced activity of the MAO can be attributed neither as an effect of the illness nor to the influence of treatment measures (e.g. neuroleptics). An association between the reduced enzyme activity and a family history of schizophrenia, as Baron and Lewitt (1980) reported, was not present in the study. The discussion of whether MAO activity is appropriate as a genetic marker for schizophrenia and to what extent it is meaningful in a pathophysiological sense cannot take place at this point. What does appear to be significant is that the twin studies provide clear indications that the level of MAO activity is controlled strictly on a genetic basis.

Pollin (1972) reported on increased urinary excretion of the catecholamines dopamine, epinephrine and norepinephrine in monozygotic twin pairs who were discordant for schizophrenia compared with control persons. This study also found a significant intrapair correlation in the urine levels of these catecholamines in the discordant pairs, i.e. both the ill and the healthy twins had higher levels. The authors integrated these findings into a pathogenetic stress model of schizophrenia in which genetic and biochemical factors and stress-inducing experiences are effective when taken together.

Wyatt et al. (1973b) investigated blood platelets (not dialyzed) of 14 monozygotic twin pairs who were discordant for schizophrenia, for their ability to produce the hallucinogen dimethyltryptamine (DMT) enzymatically. The schizophrenic index-twins showed higher enzyme levels in their serum than their healthy co-twins. The serum levels of the enzyme in the co-twins corresponded to the serum levels of 22 healthy control persons. The authors believed this to be an indication that the higher enzyme level of the schizophrenic index-twins were exclusively due to environmental influences and were not genetically determined. Putten et al. (1996) found significantly different samples of certain plasma proteins for schizophrenia-discordant monozygotic twins in gel-electrophoresis tests. In the monozygotic control twins, however, hardly any differences could be recognized. Recently Poltorak et al. (1997) reported that schizophrenic subjects of discordant monozygotic pairs showed significantly higher levels of adhesion molecules (neural cell adhesion molecule = N CAM) in the cerebrospinal fluid and lower levels of L1 antigen than their healthy partners. The healthy partners were not distinguishable from the control persons. The authors believed this to be an indication that the changes of N CAM were either dependent on influences preceding the onset of the disorder, or could be traced to influences directly affected by the disorder. In the authors' opinion genetic causes are not accountable for these changes.

Other findings. In a twins sample of 11 monozygotic and for schizophrenia discordant pairs done by Pollin et al. (1966), a clinically relevant thyroid disorder was found with unexpected frequency among the mothers (7 of 11 mothers). Since the husbands of these women had no thyroid problems, geographical artifacts were rather unlikely.

Goldberg et al. (1990) conducted a series of neuropsychological tests with 16 monozygotic pairs discordant for schizophrenia. The schizophrenic twins showed deficiencies, especially in the vigilance and memory test results, as well as in the forming of concepts compared to the healthy co-twins. These neuropsychological disorders were viewed as being independent of genetic and nonspecific environmental influences and were evaluated as being an expression of the clinical disease process itself. Unfortunately the authors provided no information on the phase of the disorder of these test persons (acute? residual? chronic? remittent?) at the time of the investigations.

Pre-psychotic personality differences

A series of twin studies ascertained that among discordant pairs, the schizophrenic partner had significantly more often been the submissive, more reserved, less self-sufficient partner, with more mental problems during childhood. Before the onset of the disease, this twin often had fewer friends of either sex and a lower social status (Essen-Möller 1941, Slater 1953, Kurihara 1959, Tienari 1968, Pollin et al. 1966, Stabenau and Pollin 1967, Kringlen 1967). Kringlen (1990) emphasizes these differences in discordant monozygotic as well as in dizygotic pairs. The most distinct difference is observed in the dominant-subordinate axis. In a survey of 100 discordant pairs from six twin studies (Slater 1953, Inouye 1961, Tienari 1963, Gottesman and Shields 1966, Kringlen 1967, Fischer et al. 1969) the more subordinate partner was subsequently the schizophrenic index-twin in no fewer than 84 cases (Gottesman and Shields 1982).

Children of discordant monozygotic twins

In 1971 Fischer reported that the children of schizophrenic index-twins and their healthy co-twins had the same risk of later developing schizophrenia. Eleven schizophrenic index-twins had 47 children, six of whom developed schizophrenia (15.5%, adjusted for age). Six healthy co-twins had 24 children, four of whom developed schizophrenia (17.4%, adjusted for age). Ten schizophrenic dizygotic index-twins had 27 children, four of whom developed schizophrenia (18.0%, adjusted for age). Twenty healthy dizygotic co-twins had 52 children, one of whom developed schizophrenia (2.7%, adjusted for age). It was possible to examine all the children eighteen years later (Gottesman and Bertelsen 1989). Fischer's findings (1971) confirm this. Kringlen and Cramer (1989), however, could not confirm these findings in their twins sample. They found that five of 28 children of monozygotic schizophrenic index-twins (17.9%) were diagnosed with disorders included in the schizophrenic spectrum, but that only two of 45 children of the healthy co-twins (4.4%) developed such disorders. The difference was not significant. Kringlen and Cramers's findings of dizygotic twin studies reflected those of Fischer. Four of 22 children of schizophrenic persons (18.2%) – in contrast to only one of 37 children of healthy persons (2.7%) – developed schizophrenia.

Leonhard's findings on twins

For clarification, a few basic remarks on Leonhard's nosological differentiation of psychoses of the schizophrenic spectrum are presented at this point. Leonhard divides psychoses into three major groups based on prognostic, psychopathological and course-specific criteria: cycloid psychosis and unsystematic and systematic schizophrenias.

In cycloid psychosis Leonhard discovered a prognostically favorable long-term course of the illness, similar to manic-depressive illness. However, in contrast to the manic-depressive illness he found only a low familial loading (Leonhard 1986, 1995). He distinguished three clinical subgroups: anxiety-happiness psycho-sis, confusion psychosis and motility psychosis. Recent investiga-tions have been able to confirm and validate the concept of cycloid psychoses (Perris 1975, Brockington et al. 1982, Beckmann et al. 1990, Strik et al. 1993,1996).

Based on specific symptomatological similarities, Leonhard describes unsystematic schizophrenia as the malevolent cousin of cycloid psychosis. The former most often appears intermit-tently and is (partially) remittent. In the long-term, mental resi-dues of varying severity will result. In unsystematic schizophrenia, Leonhard (1975) found a high familial loading of similar illnesses and therefore assumed a primarily genetic origin. Clinically he distinguished three subgroups: periodic catatonia, affective para-phrenia and cataphasia.

Systematic schizophrenia usually begins slowly, appears chro-nically without remissions and always leads to severe, lasting mental defects. Clinically he distinguished three subgroups: hebe-phrenia, systematic catatonia and systematic paraphrenia. The fa-milial loading is very low in the case of systematic schizophrenia and differs very little from the morbidity rate of the general popu-lation. Recent methodologically reliable investigations have con-firmed the different familial loading in unsystematic and system-atic schizophrenias (Beckmann et al. 1992, 1996, Franzek and Beckmann 1991, 1992a, Franzek et al. 1995, Stöber et al. 1995).

Leonhard did not conduct a systematic twin survey and therefore a statistical analysis was also not carried out. In 1978 he reported on 72 twin pairs which had been under his observation by that time. Of these pairs, 33 were monozygotic and 36 were dizygotic. Zygosity could not be determined in three of the pairs. The most significant findings were that when his differentiated nosology was applied, no monozygotic twins were found with systematic schizophrenia, whereas this psychosis did occur among the dizy-gotic twins. Up to 1986 he examined 69 monozygotic single twins with endogenous psychoses, from a total of 45 pairs. He was not able to find systematic schizophrenia among either the 45 index-twins or among the 24 ill partners! After excluding 13 pairs with phasic diseases (monopolar phasic, bipolar phasic = manic-de-pressive), the remaining cases were either unsystematic schizo-phrenia or cycloid psychoses. Among 42 dizygotic index-twins, however, twelve developed systematic schizophrenia. Leonhard (1979, 1986) posed the important question of whether the special twin constellation of most monozygotic twins could hinder the de-velopment of systematic schizophrenia.

Leonhard conducted concordance/discordance analyses only to a very limited degree. More precise information on psychoses within the "schizophrenic spectrum" can be found only in periodic catatonia (a subgroup of unsystematic schizophrenia), motility psychosis and anxiety-happiness psychosis (subgroup of cycloid psychosis).

Of six monozygotic pairs with a periodic catatonic index-twin, five pairs were concordant (83%). In the concordant pairs the first-born was always more severely ill. In one case the more severely ill twin was dominant over the partner, while neither dominated in the other four cases. In the discordant pair the ill twin was the second-born and weighed less at birth, but dominated over the partner.

Among the cycloid psychoses, nine of 11 monozygotic pairs with motility psychoses were concordant (82%). Of the two discordant pairs the ill index-twin was the second-born, more susceptible to mental strain and in one case suffered from severe asphyxia during birth. Among the concordant pairs no essential differences could be detected between the two partners based on the birth histories. In many cases, however, either one or both partners had suffered birth complications. In five of nine cases the second-born and mentally more unstable twin became ill earlier and/or more severely. Eleven dizygotic index-twins with motility psychoses all had a healthy partner.

Among six monozygotic pairs the index-twin suffered from anxiety-happiness psychoses (another subgroup of cycloid psychosis). Only one pair was concordant (17%), where the second-born and mentally more unstable twin was more severely ill. Of the five discordant pairs the second-born and/or mentally more unstable was always the ill twin. Four dizygotic pairs were discordant.

The high concordance rate of monozygotic twins with motility psychoses (82%) is surprising and provides a stark contrast to the low familial loading of these psychoses. Leonhard (1976) saw here an exception to Galton's law, which states that concordance among monozygotic twins speaks for heredity and discordance for environmental factors. When discussing the etiology of motility psychosis, he gave in addition to genetic causes, somatic causes in particular in the peri- and postnatal period, which always influenced both twins.

Further questions

Despite the enormous progress which has been made over the past several years in neuro-imaging, neurochemistry, neuropathology, molecular genetics, and epidemiological and social-psychiatric research, the understanding of the causes of psychoses within the schizophrenic spectrum has stagnated. There are hardly any findings which are accepted universally. Contradictory results are generally explained as being due to genetic heterogeneity and polygenic heredity in relationship to manifold environmental factors. The question keeps coming up: Is the spectrum of schizophrenic and schizophrenia-like psychoses a continuum of illnesses with continuous boundaries and common causes, or does it consist of separate illnesses with very different causes?

This unanswered question justifies the carrying out of a new systematic twin study which simultaneously takes into account and compares the various diagnostic criteria. In a polydiagnostic approach, the internationally applied operationalized diagnostic systems of DSM-III-R and ICD 10 were compared directly with the nosology according to Leonhard, based on empirical investigation in relation to specific variables. The Leonhard classification is based on highly differentiated descriptions of illnesses. Diagnosis is permitted only when either all, or at least the characteristic symptoms of a clinical picture are clearly present. This is the essential difference to the operationalized diagnosis systems in which a certain number of symptoms must correspond, though not all, nor must at least the characteristic symptoms within a symptom cluster be present.

The following different variables were selected: concordance/discordance in the comparison of monozygotic and dizygotic twins, familial loading among first degree relatives, number and severity of birth complications in inter- and intrapair comparison, the influence of birth order and position within the pair (dominant/subordinate) in the inter- and intrapair comparison, and the handedness of the index- and co-twins.

Methodology of a systematic twin study

Twin survey methodology and zygosity diagnosis

Twins survey. The region of Lower Franconia was selected for the survey of index-twins since this area has a stable population density with very little fluctuation due to immigration or emigration. The limited representative twin survey method was used (see page 17). All twins who were born after 1930 and were hospitalized in Lower Franconia for a mental disorder were to be recorded. For this purpose approximately 30,000 medical files in the medical records archive of the inpatient psychiatric facilities for the region of Lower Franconia (University Psychiatric Clinic Würzburg, Psychiatric Hospital Lohr am Main, Psychiatric Hospital Werneck) were systematically reviewed.

The question of personal data protection was discussed at length with the data protection commissioner at the University of Würzburg. There were no objections to the study and, specifically, no objections to the twin survey method. The condition was naturally met that only data should be evaluated from persons who had consented to take part in the investigation.

Zygosity diagnosis. Only same-sex pairs were selected for this investigation. According to the Weinberg estimation method among same-sex twin pairs which are studied within a systematic survey, approximately half can be expected to be monozygotic and half dizygotic twin pairs. The determination of genetic polymorphisms (blood group characteristics or DNA polymorphisms) is currently viewed as the most reliable method for diagnosing zygosity. For this study, zygosity was determined based on molecular-genetic methods, using highly polymorphous microsatellites and carried out by the Humangenetische Institut, Bonn, Prof. Propping (Director) and the Institut für Rechtsmedizin of the University of Würzburg, Prof. Patzelt (Director). This method has proven to be very reliable, quick and cost-effective (Erdmann et al. 1993). A zygosity questionnaire developed by Torgersen (1979) was also used (see page 15).

Methodology of psychiatric diagnostics

The psychiatric diagnoses of the index- and co-twins were compiled by two experienced psychiatrists (H.B., E.F.). The index-twins were diagnosed by H.B. while the co-twins were diagnosed by E.F. In several reliability studies, the two diagnosticians obtained a high rate of correspondence both when using the DSM-III-R as well as when using the Leonhard classification system (Franzek and Beckmann 1991, Stöber et al. 1995, Pfuhlmann et al. 1997). The index-twins were described to H.B. on the basis of existing documents (medical documents, reports, etc.). The subject was then examined extensively. The psychopathology in cross- and longitudinal section was documented by an extensive catamnesis (including the current examination), as well as by means of the structured interview guideline SADS-LA (Schedule for Affective Disorders and Schizophrenia, Lifetime Version by R.L. Spitzer and J. Endicott). The diagnoses were made according to DSM-III-R, ICD 10 and the Leonhard classification system. At the time of diagnosis of the index-twin H.B. had no information on the psychiatric diagnosis of the co-twin, nor of the zygosity of the twin pairs. H.B. was not able to make a final diagnosis of four index-twins after the initial extensive examination. After obtaining additional information about the family environment of the subject and after extensive discussion with E.F., H.B. was then able to make a clear diagnosis.

The diagnosis of the co-twin was compiled by E.F. based on personal examination and all available medical documents, reports, etc. The psychopathology and other diagnostically relevant data were also documented with a catamnesis and the SADS-LA. If one of these twins also suffered from a psychiatric relevant disorder (n = 17), he was also seen personally by H.B..

Concordance was determined when the diagnoses of both partners of one pair were unambiguous. After the conclusion of the diagnostic process and the determination of concordance, the zygosity diagnoses were then made known.

These complicated procedures in diagnosing the twins virtually excluded the risk of "contamination diagnoses". This term means that the diagnosis of a co-twin would be influenced by the diagnosis of the index-twin if both partners had been assessed by the same examiner.

Determining the criteria for concordance/discordance

According to the model presented by Gottesman and Shields (1966, 1972) and Fischer et al. (1969), graduated criteria were developed for the determination of concordance/discordance. The criteria were defined separately for each classification system (DSM-III-R, ICD 10, Leonhard classification).

Criteria for concordance/discordance in DSM-III-R

The endogenous (functional) psychoses were divided into four main groups:

- 295: Schizophrenia.
- 296: Mood disorders.
- 297: Delusional (paranoid) disorders.
- 298, 295.4, 295.7: Psychotic disorders not elsewhere classified.

The term "schizophrenia" is defined very narrowly. The categories "delusional (paranoid) disorders" and "psychotic disorders not elsewhere classified" are described extensively. The latter were divided further into:

- Brief reactive psychosis.
- Schizophreniform disorder.
- Schizoaffective disorder.
- Induced psychotic disorder.
- Psychotic disorder not otherwise specified (atypical psychosis).

The diagnoses schizophrenia, brief reactive psychosis, schizophreniform disorder, schizoaffective disorder, induced psychotic disorder and psychotic disorder not otherwise specified (atypical psychosis) all encompass our definition of the psychoses of the schizophrenic spectrum in DSM-III-R.

Three different concordance groups (C1, C2, C3) and one group for discordance D were established.

Concordance group C1. The co-twin is/was also psychotic and fulfills the same diagnostic criteria as the index-twin for the same three-digit category (295, 297, 298).

Concordance group C2. The co-twin is/was also psychotic, does/did not suffer from the same but from a different psychosis within the schizophrenic spectrum.

Concordance group C3. The co-twin does/did not suffer from a psychosis within the schizophrenic spectrum, but does/did suffer from another relevant psychiatric disorder with the exception of a demential disorder or mental retardation.

Discordance D. The co-twin was never mentally ill and proved to be mentally normal and healthy during the investigation, including the medical history.

Criteria for concordance/discordance in ICD 10

The endogenous psychoses were grouped into two main categories:

– F2: Schizophrenia, schizotype and delusional disorders.
– F3: Affective disorders.

Category F2 is divided into:

– F20: Schizophrenia.
– F21: Schizotype disorder.
– F22: Continuous delusional disorder.
– F23: Transient psychotic disorder.
– F24: Induced delusional disorder.
– F25: Schizoaffective disorder.
– F28: Other non-organic psychotic disorder.
– F29: Non-organic psychosis not otherwise specified.

The categories F20, F21, F22, F23, F24, F25, F28, F29 all encompass our definition of the psychoses of the schizophrenic spectrum in ICD 10.

Three different groups for concordance (C1, C2, C3) and one group for discordance D were established.

Concordance group C1. The co-twin is/was also psychotic and fulfills the diagnostic criteria for the same three-digit (Fxx) category as the index-twin.

Concordance group C2. The co-twin is/was also psychotic. The diagnosis is not identical to that of the index-twin, but fulfills the diagnostic criteria for one of the diagnoses from the schizophrenic spectrum.

Concordance group C3. The co-twin does/did not suffer from a schizophrenic spectrum psychosis, but does exhibit another relevant psychiatric diagnosis, with the exception of a demential disorder and mental retardation.

Discordance D. The co-twin was never psychotic and proved to be mentally normal and healthy during the investigation, including the medical history.

Criteria for concordance/discordance in the Leonhard classification

Leonhard divided the endogenous psychoses into five main categories based on the cross-sectional picture, course and outcome of the disease (see page 3, Fig.1).

1. Monopolar phasic psychoses.
2. Bipolar phasic psychoses.
3. Cycloid psychoses.
4. Unsystematic schizophrenias.
5. Systematic schizophrenias.

These five main categories are classified further into subgroups based on specific clinical pictures. For the purpose of our investigation, however, this is irrelevant for the time being. Psychoses in which so-called "schizophrenic symptoms" occur are cycloid psychoses and unsystematic and systematic schizophrenias. These three groups all encompass our definition of the psychoses of the schizophrenic spectrum in the Leonhard classification.

Three different concordance groups (C1, C2, C3) and one group for discordance D were established.

Concordance group C1. The co-twin is/was also psychotic and show/showed a psychopathological cross-sectional picture with consideration of the course of the illness, requiring assignment to the same diagnostic main category of 3 – 5 to which the index-twin had been classified.

Concordance group C2. The co-twin suffers/suffered from another psychosis of the schizophrenic spectrum.

Concordance group C3. The co-twin does not/did not suffer from a schizophrenic spectrum psychosis (in this case: cycloid psychosis, unsystematic or systematic schizophrenias), but does exhibit another relevant psychiatric disorder, except a demential disorder or a mental retardation.

Discordance D. The co-twin was never psychotic and proved to be mentally normal and healthy during the investigation, including the medical history.

The concordance rates were calculated using the pair-wise and proband-wise method. The pair-wise method of calculation was used for statistical evaluation of the concordance rates of the monozygotic and dizygotic twin pairs.

Table 9. Fuchs - rating - scale of complications during pregnancy and birth (Parnas, et al. 1982)

Severity of complication from 1 to 4	Type of complication
0 =	no complications
1 =	forceps delivery cesarean section placental defects previous miscarriages bleeding after birth adiposity of mother narrow pelvis illness of mother during pregnancy (twin birth) birth process > 24 hours
2 =	severe illness in mother placental infarcts abnormal birth position premature amniorrhexis pelvic contraction during birth primary uterine inertia signs of immaturity (birth weight > 2500g)
3 =	secondary uterine inertia bleeding during birth birth process > 48 hours
4 =	asphyxia umbilical cord twisting other umbilical cord complications eclampsia signs of immaturity(birth weight < 2500g)

Selection of further test variables

Family history. To survey the information contained in family histories, the medical documents of the index-twins were consulted first. A further survey was then based on the Family History Method. In addition to the two twin subjects, all living mothers (81%) and fathers (44%) were also examined. Mentally ill family members were investigated personally using a semi-structured interview (SADS-LA). For family members who had been treated as inpatients, medical documents were obtained after receiving official consent. This also applied for all deceased family members who had been ill. In addition to mental disorders and suicides in relatives, other long-term symptoms relating to mental illness such as personality disorders and alcohol abuse were also recorded.

Birth histories (birth order, number and severity of birth complications). The birth histories were taken from various sources. The medical records were reviewed and then the twins were interviewed. Extensive retrospective pregnancy and birth histories were taken from all living mothers. This is viewed in studies as a valid and very reliable method (Wenar and Coulter 1962, Feinleig 1985, Little 1986, Gayle, et al. 1988, O'Callaghan, et al. 1990a). Unfortunately it was not possible to obtain birth records from the clinics where the mothers had given birth. In almost every case these records had already been destroyed. Data on complications during pregnancy and birth were then rated using the internationally recognized scale "Severity Weight Allocation Scale for Specific Complications" (Table 9, Parnas et al. 1982) on the basis of number and severity.

Role assignment in the pair constellation (dominant/subordinate). The same sources of information were used for this component as for the survey of birth histories.

Handedness. A modified handedness questionnaire according to Annett (1970) was given to all twins to be answered personally. The test persons were divided into exclusively right-handed, ambidextrous and exclusively left-handed groups. The ambidextrous and exclusively left-handed groups were combined and designated as "non-right-handed".

Results

Results of the twin survey and zygosity diagnosis

Twin survey. A total of 452 index-twins were found in the approximately 30,000 medical records of the psychiatric hospitals in the region of Lower Franconia (Lohr and Werneck psychiatric hospitals, University Clinic of Würzburg), i.e. approximately one in every 66 patients could be identified as a twin birth. This is more or less in line with the frequency of twins in the general population so that no significant number of twins receiving psychiatric inpatient treatment was overlooked. According to the medical records, the twin's partner had already died in 82 cases. Of the remaining 370 index-twins with a surviving co-twin at the time of hospitalization, 234 pairs were of the same sex and 136 pairs were of the opposite sex. According to the diagnosis in the medical records, 121 of the same-sex index-twins suffered from endogenous psychoses, and 113 index-twins had other diagnoses (addiction, neurosis, organic brain disease etc.). According to the medical records, 77 index-twins from 66 pairs had ICD 9 diagnoses that belong to the category of the schizophrenic spectrum. One partner was already deceased at the time of the study in six of the 66 pairs. In eight of the remaining 60 pairs (13%), one or both partners refused to participate in the study. The personal examination was unable to confirm the diagnosis of a psychosis of the schizophrenic spectrum in five pairs. *The study is therefore based on a total of 47 same-sex twin pairs with at least one index-twin suffering from a psychosis of the schizophrenic spectrum.*

Zygosity. Blood samples for a molecular-genetic zygosity diagnosis could be taken from both partners in the case of 43 pairs. The dizygotic/monozygotic diagnoses of these 43 pairs were identical with those of the questionnaire method in 42 cases (97.6%). No blood samples could be obtained in the case of three pairs and the blood taken from one pair was not sufficient. The zygosity diagnoses in these four cases are based solely on the questionnaire method and the similarity test. All the zygosity diagnoses are listed with the respective degree of probability in Table A1 of the Appendix. There were 22 monozygotic and 25 dizygotic twin pairs.

Psychiatric diagnoses

In Tables A2, A3, A4 and A5 of the Appendix the diagnoses of all subjects are listed separately according to the operationalized classifications (DSM-III-R/ICD 10) and the nosological classification (Leonhard). A total of 64 subjects from 47 pairs suffered from psychoses of the schizophrenic spectrum.

According to the *DSM-III-R criteria*, the 64 ill subjects were distributed among 30 schizophrenias, 17 schizoaffective disorders, five delusional disorders, eight schizophreniform disorders and four atypical psychoses.

When applying the *ICD 10 criteria* there were 30 schizophrenias, 15 schizoaffective disorders, five delusional disorders, 11

Table 10. Distribution of subjects classified according to Leonhard among the various diagnoses in the operationalized classifications (DSM-III-R/ICD 10)

DSM-III-R		ICD 10	
Unsystematic schizophrenia *n = 31*			
Schizophrenia	n = 20	schizophrenia	n = 20
Schizoaffective dis.	n = 4	schizoaffective dis.	n = 4
Schizophreniform dis.	n = 1	acute transient psychot. dis.	n = 1
Delusional disorder	n = 4	delusional disorder	n = 4
Atypical psychosis	n = 2	other non-organic psychotic disorders	n = 2
Systematic schizophrenia *n = 6*			
Schizophrenia	n = 6	schizophrenia	n = 6
Cycloid psychoses *n = 27*			
Schizophrenia	n = 4	schizophrenia	n = 4
Schizoaffective dis.	n = 13	schizoaffective dis.	n = 11
Schizophrenif. dis.	n = 7	acute transient	
Atypical psychosis	n = 2	psychotic disorder	n = 10
delusional disorder	n = 1	delusional disorder	n = 1
		non-organic psychosis not otherwise specified	n = 1

acute transient psychotic disorders and three other non-organic psychotic disorders.

According to the *Leonhard classification*, 31 unsystematic schizophrenias, six systematic schizophrenias and 27 cycloid psychoses were diagnosed.

Table 10 shows how the subjects diagnosed according to Leonhard are distributed among the various DSM-III-R and ICD 10 diagnoses. All the systematic schizophrenia also met the criteria for schizophrenia according to DSM-III-R and ICD 10. In comparison, both the unsystematic schizophrenia and the cycloid psychoses were distributed over a variety of DSM-III-R/ICD 10 diagnoses.

No systematic schizophrenia was found in the case of the monozygotic subjects, whereas this illness accounted for almost one third of all diagnoses in the case of the dizygotic subjects.

The number in the subgroups of the 31 unsystematic schizophrenia varied widely. A cataphasia was only diagnosed twice (both twins of a monozygotic pair), whereas 13 subjects suffered from an affective paraphrenia and 16 subjects from a periodic catatonia.

Demographic data of the subjects

Age, age at onset and duration of illness

The subjects of the 47 pairs were 40 years old at the time of the study (± 13 standard deviation, range: 22 to 65 years). The average age of the monozygotic twins (22 pairs) was 41 years (± 12 standard deviation, range: 22 to 63 years) and that of the dizygotic twins (25 pairs) was 39 years (± 13 standard deviation, range: 22 to 65 years). The 22 female pairs were on average significantly older than the 25 male pairs (45 years ± 13 standard deviation versus 34 years ± 11 standard deviation, t-test for unpaired data: t = 2,538; p < 0.05). These age differences were also evident even when monozygotic female pairs were compared with monozygotic male pairs and when dizygotic female pairs were compared with dizygotic male pairs (Table 11).

The average age at the onset of the illness of the ill index- and co-twins (n = 64) was 21 years (± 9 SD). On average there was an interval of 19 years between the onset and the follow-up study of the ill twins (± 13 SD, range: 2 years to 45 years). The onset of the illness was defined by the first objectively reported and documented symptoms of the illness. This was generally identical with the time of the first hospitalization.

Table 12 shows that in 94% of the cases the interval between the onset and the follow-up study of ill subjects was more than four

Table 11. Age differences between female and male twin pairs

Subjects	Average age at the time of the study
Female pairs: (n = 22)	45 years (± 13 SD)
	p < 0.05[1]
Male pairs: (n = 25)	34 years (± 11 SD)
Female monozygotic pairs: (n = 10)	47 years (± 11 SD)
	p < 0.05[2]
Male monozygotic pairs: (n = 12)	37 years (± 12 SD)
Female dizygotic pairs: (n = 12)	43 years (± 15 SD)
	ns[2]
Male dizygotic pairs: (n = 13)	34 years (± 10 SD)

[1] T-test for unpaired data (t = 2.538).
[2] U-test (small random sample).

years, in 64% of the cases it was more than nine years and in 55% of the cases it was more than 14 years.

Table 13 shows that there were slight age differences in the age at the onset between men and women within individual diagnostic

Table 12. Interval between the time of the onset and the follow-up study of the 64 ill subjects

Interval	Number (percentage) of subjects
0 to 4 years	4 (6.3%)
5 to 9 years	19 (29.7%)
10 to 14 years	6 (9.4%)
15 to 19 years	9 (14.1%)
20 to 24 years	3 (4.6%)
25 to 29 years	6 (9.4%)
30 to 34 years	6 (9.4%)
35 to 39 years	7 (10.9%)
40 to 45 years	4 (6.3%)

Table 13. Age at the onset in the different diagnostic categories comparing men and women in each case

Diagnoses	Age at onset female twins	male twins
Schizophrenic spectrum	(n = 31) 22.5 years (± 9.8 SD)	(n = 33) 19.3 years (± 6.5 SD)
Schizophrenia (DSM-III-R/ICD 10)	(n = 11) 16.4 years (± 6.9 SD)	(n = 17) 16.6 years (± 7.5 SD)
Other diagnoses of the schizophrenic spectrum (DSM-III-R/ICD 10)	(n = 20) 24.9 years (± 10.6 SD)	(n = 16) 21.1 years (± 3.9 SD)
Systematic schizophrenia	(n = 1) 24 years	(n = 5) 17 years (± 8 SD)
Unsystematic schizophrenia	(n = 15) 18.4 years (± 6.7 SD)	(n = 16) 19.5 years (± 7.9 SD)
Cycloid psychoses	(n = 15) 26.5 years (± 10.8 SD)	(n = 12) 24.1 years (± 7.2 SD)

subgroups (cycloid psychoses and other spectrum psychoses apart from schizophrenia according to DSM-III-R/ICD 10).

Social situation (school, occupation, marital status)

Tables A6 and A7 in the Appendix show the social situation of the subjects at the time of the first hospitalization and at the time of the follow-up study.

In 30 of the 47 twin pairs only one twin partner had a psychotic disorder (= 30 discordant pairs: 10 monozygotic and 20 dizygotic pairs). In 17 pairs both partners were ill (= 17 concordant pairs: 12 monozygotic and five dizygotic pairs).

Of the 12 monozygotic concordant pairs there were 11 cases (92%) in which both partners had the same education. There was only one pair (8%) in which each twin had a different education. Of the 10 discordant monozygotic pairs there were seven cases (70%) in which both partners went to the same school and three cases (30%) in which the healthy partner had a higher level of education.

Five dizygotic pairs were concordant. In this case each of the two partners of a pair had the same education. Of the 20 dizygotic discordant pairs there was one case of the ill partner having a higher, and four cases of the ill partner having a lower level of

education. In 15 of the dizygotic discordant pairs the education
of both partners was the same. *The frequency of the ill partner
having a lower level of education than the healthy partner was
therefore more or less the same for discordant monozygotic and
discordant dizygotic pairs.* Overall in the case of discordant pairs
there were seven cases in which the twin who became ill later had
a lower level of education, and one case in which he had a higher
level of education. In 22 cases both partners had the same education. The findings are not statistically significant.

Table 14 shows that the education of the 30 ill twins from the
discordant pairs was no different from the education of the 34 ill
twins from the 17 concordant pairs.

Table 15 shows the social situation of the ill subjects at the time
of the follow-up study. The 34 subjects of the 17 concordant pairs
(12 monozygotic, five dizygotic pairs) are compared with the 30
subjects of the 30 discordant pairs (10 monozygotic, 20 dizygotic
pairs). It is evident that the social situation of the ill twins of the
discordant pairs was not different from the social situation of the ill
twins of the concordant pairs.

All 30 healthy partners of the discordant pairs were employed
and/or integrated in a family at the time of the follow-up study (23
employed, seven homemakers). This was the case in only 10 of the
30 ill partners (33%) (seven employed, three homemakers). 25 of
the 30 healthy partners (83%), but only six of the 30 ill partners
(20%) were married or living together with a partner.

Table 16 shows the education of the ill subjects based on the differentiated diagnoses according to Leonhard. It is not appropriate
to divide this table into monozygotic and dizygotic subjects due to
the low number of cases which would then result. It is clear that
subjects with periodic catatonia left school without any qualifica-

Table 14. Educational comparison of the 34 ill twins from the 17 concordant pairs (12
monozygotic, 5 dizygotic pairs) with the 30 ill twins from the discordant pairs (10 monozygotic, 20 dizygotic pairs)

	No schooling completed/ school for teaching disabled children	Middle school	Higher-level school
Ill twins of the discordant pairs (n = 30)	6 (20%)	14 (47%)	10 (33%)
Ill twins of the concordant pairs (n = 34)	12 (35%)	12 (35%)	10 (29%)

Table 15. Social situation comparing the 34 ill twins from the 17 concordant pairs (12 monozygotic, 5 dizygotic pairs) with the 30 ill twins from the discordant pairs (10 monozygotic, 20 dizygotic pairs) at the time of the follow-up study

Social situation	Subjects of concordant pairs (n = 34)	Ill subjects of discordant pairs (n = 30)
Employed/homemaker/ normal retirement	9 (26%)	10 (33%)
Unemployed/rehabilitation	10 (29%)	9 (30%)
Early retirement	4 (12%)	5 (17%)
Homeless	3 (9%)	0
Nursing home/ permanently hospitalized	8 (24%)	6 (20%)
Married/ living together	7 (21%)	6 (20%)

Table 16. Education of the 64 ill twins (monozygotic and dizygotic) according to differentiated diagnoses

Diagnoses	No schooling completed/ school for teaching disabled children	Middle school	Higher-level school
Twins with systematic schizophrenia (n = 6)	2 (33%)	1 (17%)	3 (50%)
Twins with cataphasia (n = 2)	2	0	0
Twins with periodic catatonia (n = 16)	11 (69%)	5 (36%)	0
Twins with affective paraphrenia (n = 13)	0	4 (36%)	9 (69%)
Twins with cycloid psychoses (n = 27)	3 (10%)	16 (59%)	8 (30%)

tions or attended a special school more often than all the other subjects.

There were also considerable differences between the individual diagnostic subgroups in the social outcome (Table 17). Of the subjects with systematic schizophrenia, none was socially integrated at the time of the first hospitalization, which was also the case during the follow-up study. The same was true of the subjects with periodic catatonia. Approximately half were still employed shortly before their first hospitalization but none was employed at the time of the follow-up study. Of the subjects with affective paraphrenia and cycloid psychoses, 100% were socially integrated at the time of the first hospitalization. At the time of the follow-up study, 67% of the subjects with cycloid psychoses and 38% of the subjects with affective paraphrenia were still socially well integrated.

Table 17. Social situation of the 64 ill twins (monozygotic and dizygotic together) at the time of the first hospitalization and the follow-up study according to differentiated diagnoses

Diagnoses	Employed/housewife student/retired		Married/ living together	
	First hosp./follow-up study		First hosp./follow-up study	
Twins with systematic schizophrenia (n = 6)	0	0	0	0
Twins with cataphasia (n = 2)	2	0	0	0
Twins with periodic catatonia (n = 16)	9 (56%)	0	0	0
Twins with affective paraphrenia (n = 13)	13 (100%)	5 (38%)	3 (23%)	4[1](31%)
Twins with cycloid psychoses (n = 27)	27 (100%)	18 (67%)	6 (22%)	8[2](30%)

[1] One subject got married after the first hospitalization, but was divorced at the time of the follow-up study.
[2] One subject who was married at the time of the first hospitalization was divorced at the time of the follow-up study. One subject who was single at the time of the first hospitalization was widowed by the time of the follow-up study.

Table 18. Interval until the onset of concordance in the 17 pairs which were concordant at the time of the study

Pair-code	Zygosity	Interval (years)	Diagnoses
W1	MZ	0	affective paraphrenia
M2	MZ	1	cycloid psychosis
M8	MZ	0	periodic catatonia
W9	MZ	0	periodic catatonia
M12	DZ	2	periodic catatonia
M15	MZ	2	periodic catatonia
M19	MZ	1	periodic catatonia
W21	DZ	4	cycloid psychosis
W27	MZ	8	cycloid psychosis
W29	MZ	0	cycloid psychosis
W32	MZ	4	periodic catatonia
W36	MZ	16	affective paraphrenia
W38	MZ	8	affective paraphrenia
M40	DZ	4	cycloid psychosis
M41	DZ	7	affective paraphrenia
M42	MZ	2	cataphasia
W43	DZ	2	periodic catatonia

Concordance rates

In more than 90% of our subjects, the onset of the initial disorder of the index-twin took place more than four years earlier (Table 12). On average, there was an interval of 3.9 years ($\pm$ 5.1 SD) in the 17 concordant pairs until the onset of concordance.

In 76% of the cases, the pairs became concordant within four years. The interval was more than 10 years in only one case (Table 18). All the subjects with periodic catatonia became concordant within four years. These subjects were significantly different ($x^2 = 6,2$, df = 2, p < 0.05) from the subjects with affective paraphrenia, of whom two pairs did not become concordant until much later (eight and 16 years later). One pair became concordant for a cycloid psychosis after eight years. Table 19 shows the reference rates calculated with age adjustment according to Strömgren (see page 18) for the subjects that were healthy at the time of the study.

Pair-wise calculation of the concordance rates (see pages 15, 16)

The terms *C1, C2 and C3 represent the degree of concordance of the co-twin to the index-twin* (see pages 38–40). Precise definitions were given for concordance in each classification system. In the operationalized diagnosis systems (DSM-III-R and ICD 10), Group

Table 19. Reference rates for age adjustment according to Strömgren (1936, from original table) for the subjects that were healthy at the time of the follow-up study (0.00 = the subject is not counted; 1.00 = the subject is fully counted)

Subject no.	Age at time of investigation	Age at onset of disorder of the partner	Reference rate
M 3–1	42 yrs.	14 yrs.	1.00
F 4–2	45 yrs.	18 yrs.	0.96
M 6–1	53 yrs.	28 yrs.	1.00
M 7–1	29 yrs.	22 yrs.	0.73
F 10–2	50 yrs.	18 yrs.	0.99
M 11–1	28 yrs.	26 yrs.	0.68
M 13–2	41 yrs.	5 yrs.	1.00
M 14–2	22 yrs.	19 yrs.	0.38
M 16–1	37 yrs.	18 yrs.	0.97
F 17–1	42 yrs.	24 yrs.	0.90
M 18–1	25 yrs.	17 yrs.	0.66
F 20–1	34 yrs.	24 yrs.	0.74
F 22–1	22 yrs.	14 yrs.	0.64
F 23–2	31 yrs.	26 yrs.	0.63
F 24–1	42 yrs.	27 yrs.	0.89
M 25–2	57 yrs.	28 yrs.	1.00
F 26–3	33 yrs.	18 yrs.	0.80
M 28–1	33 yrs.	27 yrs.	0.83
M 30–1	25 yrs.	17 yrs.	0.66
F 31–1	46 yrs.	28 yrs.	0.67
F 33–1	64 yrs.	19 yrs.	1.00
M 34–2	32 yrs.	19 yrs.	0.95
F 35–2	55 yrs.	21 yrs.	0.96
M 37–1	59 yrs.	15 yrs.	1.00
F 39–2	27 yrs.	25 yrs.	0.59
F 44–1	34 yrs.	27 yrs.	0.83
F 45–2	62 yrs.	42 yrs.	1.00
M 46–2	29 yrs.	24 yrs.	0.71
M 47–2	36 yrs.	30 yrs.	0.79

C1 is defined as being that where both subjects of a pair suffer from a psychosis with the same three-figure code (strict concordance criterion). Group C1 + C2 means that both subjects suffer from some psychosis of the schizophrenic spectrum (this criterion was used in most early twin studies). Group C1 + C2 + C3 combines the pairs in which both partners are (were) psychotic *and* the pairs in which the co-twin is (was) not psychotic, but another relevant mental disorder is visible (broadest concordance criterion).

The pairwise concordance rates calculated according to the operationalized diagnostics are shown in Table 20. Concordance group C1 + C2 must be considered for a direct comparison with previous twin studies in schizophrenic research. The rates are 50% for monozygotic pairs and 20% for dizygotic pairs (age-adjusted: 55% and 23%). These values coincide with the average from the concordance rates of all previous twin studies (monozygotic pairs 54%, dizygotic pairs 10%, without age adjustment).

When only the psychoses, which fulfill the strict criteria for schizophrenia are considered, there is a significantly higher concordance rate for monozygotic pairs of 78%. This more or less corresponds to the concordance rates which previous authors (Kallmann, Inouye, Gottesman and Shields) had found for "severely ill" index-twins (see Table 5). A considerably lower concordance rate of only 31% (age-adjusted 36%) was calculated for the "other diagnoses of the schizophrenic spectrum". This corresponds in Table 5 to the concordance rates reported for "mildly or temporarily ill" index-twins.

In our test group, the concordance rates of monozygotic subjects were statistically significantly higher in the overall group of the schizophrenic spectrum and in the more closely defined schizophrenic group than the concordance rates of the dizygotic subjects. The concordance rate of the monozygotic subjects in the closely defined schizophrenia group was to a statistically tendency

Table 20. Pairwise concordance rates when applying an operationalized diagnostic according to DSM-III-R/ICD 10 (C1, C2 and C3 indicate different definitions of concordance). Age-adjusted values are in parentheses

Schizophrenic spectrum			
	MZ $n = 22$	DZ $n = 25$	test for proportions
Group C1	10 = 46% (50%)	4 = 16% (18%)	
95% confidence interval	24% – 68%	1% – 33%	$z = 1.77, p = .075$
Group C1+C2	11 = 50% (55%)	5 = 20% (23%)	
95% confidence interval	28% – 72%	4% – 40%	$z = 1.65\ p = .099$
Group C1+C2+C3	14 = 64% (70%)	6 = 24% (27%)	
95% confidence interval	44% – 88%	8% – 44%	$z = 2.26\ p = .002$

Strict criteria of schizophrenia			
	MZ $n = 9$	DZ $n = 12$	test for proportions
Group C1	6 = 67% (67%)	2 = 17% (18%)	
95% confidence interval	36% – 98%	0% – 39%	$z = 1.88, p = .06$
Group C1+C2	7 = 78% (78%)	3 = 25% (27%)	
95% confidence interval	56% – 100%	1% – 50%	$z = 1.97, p = .049$
Group C1+C2+C3	9 = 100%	3 = 25% (27%)	
95% confidence interval	100%	0% – 50%	$z = 2.99, p = .003$

Other diagnoses of the schizophrenic spectrum			
	MZ $n = 13$	DZ $n = 13$	test for proportions
Group C1	4 = 31% (36%)	2 = 15% (18%)	ns
Group C1+C2	4 = 31% (36%)	2 = 15% (18%)	ns
Group C1+C2+C3	5 = 38% (45%)	3 = 23% (28%)	ns

higher than the concordance rate of the remaining spectrum psychoses according to DSM-III-R/ICD 10 ($x^2 = 3,01$, df = 1, p < .1).

The pairwise concordance rates when using the Leonhard classification is shown in Table 21. The concordance group C1 in this case means that both partners of a pair belong to the same diagnostic subgroup of the schizophrenic spectrum (unsystematic schizophrenia, systematic schizophrenia or cycloid psychosis). Concordance group C1 + C2 combines pairs in which both partners have a psychosis from the schizophrenic spectrum. Concordance group C1 + C2 + C3 in addition covers pairs in which the co-twin is (was) not psychotic, but suffers from another relevant mental disorder (see page 40).

In the case of systematic schizophrenia, there were no monozygotic twins and all six dizygotic pairs were discordant. The pairwise concordance rate of the unsystematic monozygotic twins was 82% (age-adjusted 82%) and that of the dizygotic pairs was 25% (age-adjusted 27%). In the prognostically favorable cycloid psychoses the pairwise concordance rate of monozygotic pairs was 27% (age-adjusted 33%) and that of the dizygotic pairs was 18% (age-adjusted 20%). The statistical calculations showed a significant

Table 21. Pairwise concordance rates when using the Leonhard classification (C1, C2 and C3 indicate different definitions of concordance). Age-adjusted values are in parentheses

Systematic schizophrenia			
	MZ n = 0	DZ n = 6	
Group C1	–	0%	
Group C1+C2	–	0%	
Group C1+C2+C3	–	0%	
Unsystematic schizophrenia			
	MZ n = 11	DZ n = 8	test for proportions
Group C1	9 = 82% (82%)	2 = 25% (27%)	
95% confidence interval	59% – 100%	0% – 55%	z = 2.02, p = .044
Group C1+C2	9 = 82% (82%)	2 = 25% (27%)	
95% confidence interval see above			p = .044
Group C1+C2+C3	11 = 100%	4 = 50% (53%)	
95% confidence interval	100%	15% – 85%	z = 2.07, p = .038
Cycloid psychoses			
	MZ n = 11	DZ n = 11	test for proportions
Group C1	3 = 27% (33%)	2 = 18% (22%)	ns
Group C1+C2	3 = 27% (33%)	2 = 18% (22%)	ns
Group C1+C2+C3	3 = 27% (33%)	2 = 18% (22%)	ns

difference in the concordance rates of monozygotic and dizygotic pairs only in the group of unsystematic schizophrenia. The concordance rate of the monozygotic subjects with unsystematic schizophrenia was statistically significantly higher than the concordance rate of the monozygotic subjects with cycloid psychoses ($x^2 = 5,3$, df = 1, p < .05).

Proband-wise calculation of the concordance rates

The proband-wise calculated concordance rate is directly comparable with the empirical repetition figures from family studies. The average proband-wise concordance rate of all previous twin studies is 58% for monozygotic twins and 15% for dizygotic twins. Table 22 shows that the results of our study using operationalized diagnostics in the psychoses of the schizophrenic spectrum (concordance group C1 + C2) hardly differ from this. Monozygotic subjects were concordant in 65% of cases (age-adjusted 69%) and dizygotic subjects were concordant in 26% of cases (age-adjusted 29%).

The index for hereditariness (= measure of the genetic proportion of the total variance observed for a continuously distributed characteristic) was 0.60 in our test group, and thus slightly less (not significant) than the average of all previous twin studies (0.72).

Table 22. Proband-wise concordance rates (n = number of index cases) when using an operationalized diagnostic (DSM-III-R/ICD 10). (C1, C2 and C3 indicate different definitions of concordance). Age-adjusted values are in parentheses

Schizophrenic spectrum

	MZ n = 31	DZ n = 27	index for hereditariness	MZ/DZ quotient
Group C1	19 = 61% (65%)	6 = 22% (25%)	0.64	2.77
Group C1+C2	20 = 65% (69%)	7 = 26% (29%)	0.60	2.50
Group C1+C2+C3	23 = 74% (79%)	8 = 30% (34%)	0.59	2.47

Strict criteria of schizophrenia

	MZ n = 14	DZ n = 12	index for hereditariness	MZ/DZ quotient
Group C1	11 = 79% (79%)	2 = 17% (18%)	0.78	4.65
Group C1+C2	12 = 86% (86%)	3 = 25% (27%)	0.71	3.44
Group C1+C2+C3	14 = 100%	3 = 25% (27%)	0.75	4.00

Other diagnoses of the schizophrenic spectrum

	MZ n = 17	DZ n = 15	index for hereditariness	MZ/DZ quotient
Group C1	8 = 47% (53%)	4 = 26% (29%)	0.45	1.81
Group C1+C2	8 = 47% (53%)	4 = 26% (29%)	0.45	1.81
Group C1+C2+C3	9 = 53% (60%)	5 = 33% (36%)	0.40	1.61

For index-twins who were diagnosed schizophrenic according to DSM-III-R/ICD 10, a hereditariness index of 0.71 was calculated in our group, whereas the value of the "other diagnoses of the schizophrenic spectrum" was only 0.45.

The quotient from the monozygotic and dizygotic concordance rates was over 3.5 in the group with DSM-III-R/ICD 10 schizophrenia, whereas the quotient in the group of the remaining DSM-III-R/ICD 10 diagnoses was just below 2.

The proband-wise concordance rates when using the classification according to Leonhard are shown in Table 23. In the case of monozygotic unsystematic schizophrenics, it was 89% compared with 25% for the dizygotic twins (age-adjusted 27%). The calculation of hereditariness resulted in a value of 0.72. The quotient of the concordance rates from the monozygotic and dizygotic pairs was more than 3.5.

In the case of cycloid psychoses, the concordance rate of the monozygotic twins was only slightly higher (39%, age-adjusted 45%) than that of the dizygotic twins (31%, age-adjusted 35%). The hereditariness index was therefore also correspondingly low at only 0.21. The quotient of the concordance rates from monozygotic and dizygotic pairs was only slightly more than 1.

Concordance Group C1 + C2 + C3 must be given special attention. This broad definition of the criteria for concordance between

Table 23. Proband-wise concordance rates (*n* number of index cases) when using the Leonhard classification (C1, C2 and C3 indicate different definitions of concordance). Age-adjusted values are in parentheses

Systematic schizophrenia				
	MZ n = 0	DZ n = 6	index for hereditariness	MZ/DZ quotient
Group C1	–	0%	–	–
Group C1+C2	–	0%	–	–
Group C1+C2+C3	–	0%	–	–
Unsystematic schizophrenia				
	MZ n = 18	DZ n = 8	index for hereditariness	MZ/DZ quotient
Group C1	16 = 89% (89%)	2 = 25% (27%)	0.72	3.56
Group C1+C2	16 = 89% (89%)	2 = 25% (27%)	0.72	3.56
Group C1+C2+C3	18 = 100%	4 = 50% (53%)	0.50	2.00
Cycloid psychoses				
	MZ n = 13	DZ n = 13	index for hereditariness	MZ/DZ quotient
Group C1	5 = 39% (45%)	4 = 31% (35%)	0.21	1.25
Group C1+C2	5 = 39% (45%)	4 = 31% (35%)	0.21	1.25
Group C1+C2+C3	5 = 39% (45%)	4 = 31% (35%)	0.21	1.25

index-twin and co-twin increases the concordance rates of the monozygotic as well as the dizygotic pairs in almost all diagnostic groupings, with the exception of the category of systematic schizophrenia (Tables 20, 21, 22, 23). *In the case of monozygotic pairs with schizophrenia according to DSM-III-R/ICD 10 criteria and with unsystematic schizophrenia according to Leonhard, the concordance of monozygotic twins actually increased to 100%.*

In the case of dizygotic pairs with schizophrenia according to DSM-III-R/ICD 10, concordance increased only slightly to 25%, while it increased to 50% for dizygotic pairs with unsystematic schizophrenia. This discrepancy occurs because in the case of the dizygotic pairs with DSM-III-R/ICD 10 schizophrenia, all cases of systematic schizophrenia according to Leonhard (which only occurred in dizygotic subjects) occurred, which without exception had a healthy partner.

Case studies

Monozygotic concordant pairs

Of the 22 monozygotic pairs, 14 pairs belong to the concordance groups C1, C2 and C3 (W1, M2, M8, W9, M15, M19, M25, W27, W29, W32, W36, M37, W38, M42). Their case histories, including their family medical histories, are summarized below.

Subjects W 1–1 and W 1–2 (concordance group C1)

W 1–1 and W 1–2 became ill at same time when they were 25 years old, and were admitted as inpatients on their mother's initiative. On admission, they showed symptoms of logorrhea and were very excitable and agitated. They vehemently denied having feelings of inhibition, persecution and social isolation. Even after stabilization, they lacked any insight into their illness. They felt cheated of their inheritance and believed to have been poisoned. After their discharge, they became homeless and unemployed, wandering about and sleeping in hotels at the expense of social welfare. Several inpatient stays followed. Paranoid delusions of intrusion with hostile, agitated behavior were always prevalent. *In the follow-up study,* the sisters were 35 years old, disheveled and eccentrically dressed. Logorrhea and agitated and aggressive affect were again present. They again felt they had been cheated of their inheritance and poisoned with drugs. They believed that others were trying to psychologically destroy them. The sisters believed themselves to be "special people" and entitled to special care. Hallucinations and a disturbed sense of self were absent during the whole course of the illness.

Family history. The 60 year old mother and the brother, four years older than the sisters, were personally examined. The mother had been in outpatient psychiatric care for depression for many years. During examination she was very talkative, excessive appearance and described herself as anxious and unstable. The brother of the twins appeared psychologically stable and healthy. The father was deceased and had been an alcoholic. He was considered to have been overbearing and had often been offensive. He had occasionally expressed grandiose delusions. He had sometimes spoken incoherently and felt persecuted.

Diagnoses. *W 1–1/2 according to DSM-III-R and ICD 10, delusional disorder.*

Leonhard Classification. *W 1–1/2 affective paraphrenia.*

Subjects M 2–1 and M 2–2 (concordant C1)

M 2–2 became ill when he was 21 years old, one year earlier than his brother M 2–1 did. So far, four acute psychotic episodes had occurred, each followed by a full remission. During the first phase, he asked the admitting physicians to dance with him and was skipping and prancing about. His affect was either cheerful and elevated or agitated and demanding. He was bubbling over with ideas, but was not incoherent. He expressed grandiose and omnipotent ideas. During several phases, persecutory delusions, delusions of reference and thought broadcasting were observed, in addition to the grandiose ideas. Every phase of the illness was accompanied by extreme psychomotor activity, often excited and purposeless. He once lay completely rigid for several hours in bed.

The brother *M 2–1* came to professional attention due to his dangerous behavior in traffic. In the hospital, his mood was elevated, he expressed unconnected grandiose ideas, and believed he could cause incredible things to happen. In addition, he felt observed and threatened and had sudden feelings of anxiety. He claimed to hear many different voices, including the voice of Jesus. His behavior was characterized by purposeless, psychomotor activity, he emitted repetitious non-articulated sounds and made bizarre movements. During the second phase, his extreme psychomotor agitation was very conspicuous. *In the follow-up study,* the subjects were 29 years old and healthy.

Family history. The 65 year old father, the third born triplet sister, and a four year older brother could all be examined in person. The father and the triplet sister were healthy. The older brother had twice been in inpatient care because of an endogenous depression. During examination, he was shy and reserved, but showed no odd behavior. The mother had passed away at the age of 59 years. She had been a hypochondriac for many years following a serious car accident. A three year older sister, two older brothers (six and four years older) and a sister three years younger were reported to be psychologically healthy.

Diagnoses. *M 2–1/2 according to DSM-III-R, schizoaffective psychosis, according to ICD 10, acute transient psychotic disorder.*

Leonhard Classification. *M 2–1/2 cycloid psychosis (motility psychosis with anxiety).*

Subject M 8–1 and M 8–2 (concordant C1)

Both subjects had shown repeated aggressive behavior towards others and themselves starting at the age of four, which had led to placement in a children's home. When they were 19 years old, they were admitted to a psychiatric hospital. Both showed an increase in the frequency and severity of periodic apparently purposeless and stereotyped psychomotor states of agitation, maintaining rigid postures and showing aggression. During these phases, auditory hallucinations also occurred repeatedly. Even when the subjects were affectionate and responded well to supervision in the residual phase, a discharge was not possible. Eventually the affective disturbances could be put under control only by a stereotactic operation.

In the follow-up study, the subjects were 40 years old. They did not speak at all, showed flat affect, and when addressed responded lethargically. Their psychomotor behavior was awkward, rigid and incongruous. There was no indication of delusions and hallucinations. The subjects walked around the ward holding hands.

Family history. No family member could be examined. The subjects had no siblings. Nothing was mentioned in their medical history about their deceased mother. The father was an alcoholic and had committed suicide.

Diagnoses. *M 8–1/2 according to DSM-III-R and ICD 10, catatonic type of schizophrenia.*

Leonhard Classification. *M 8–1/2 periodic catatonia.*

Subjects W 9–1 and W 9–2 (concordant C1)

The twins were placed into a children's home when they were two years old, as they were neglected by their mother. At the age of four, they were admitted to a psychiatric hospital due to their extreme agitation. The following years were marked by sequential, week-long purposeless psychomotor activity combined with unpredictable aggression and destructiveness, and quiet and affectionate behavior. Both subjects eventually went to a school for disabled children and later could be integrated into a special workplace for the disabled. At 27 years, they suddenly showed extreme behavior again. Excessive, often hysterical, inappropriate and aggressive affect occurred, with a tendency towards self-destructive acts. Over the course of several weeks negativism and dysphoric irritability alternated with agitated whining behavior. Acoustic hallucinations were assumed, and W 9–2 expressed a fear of becoming poisoned. For days at a time they exhibited intrusive and clinging social behavior. Treatment on separate wards was inevitable. In between those periods there were months when to a large extent the twins showed no severe symptoms and were amenable. *In the follow-up study,* the subjects were 34 years old. They both exhibited a rigid and passive expression and appeared sluggish and uninterested. Severe flat affect was evident. Their responses were short and vague. Delusions and hallucinations were absent.

Family history. Family members could not be examined. The mother had reportedly had four children of three different partners. She reportedly had neglected her children and was inclined to occasional excessive alcohol consumption. The father was reported to have been overbearing and lacking in self-control and also tended towards alcohol abuse. A sister of the father was said to have been "intellectually limited". A brother of the paternal grandfather was designated as a "crackpot" in his village. The circumstances of the subjects' half siblings were unknown.

Diagnoses. *W 9–1/2 according to DSM-III-R and ICD 10, catatonic type of schizophrenia.*

Leonhard Classification. *W 9–1/2 periodic catatonia.*

Subjects M 15–1 and M 15–2 (concordant C1)

The twins grew up in a children's home and visited a school for disabled children. Starting at the age of 15, *M 15–1* showed signs of recurring, severely excited psychomotor behavior, including aggression. After such states, he was often irritable for several months, resisting instructions or being intrusive. Commanding voices were also mentioned. Between the active phases, he was rather sluggish and uninterested. At the age of 30, he was discharged from the hospital and was placed in a supervised group home; however, soon thereafter was readmitted due to renewed catatonic states of agitation. *In the follow-up study,* the 41 year old subject showed flat affect, was sluggish and uninterested. Psychomotor behavior was stiff and awkward, and delusions and hallucinations were absent.

The brother *M 15–2* fell ill with a similar psychosis at the age of 17. Also, periodically recurring, severely excited psychomotor behavior occurred. During the residual phases, he also showed affective blunting without positive symptoms or was annoyingly intrusive. Discharge from the hospital was not possible. *In the follow-up study,* he was 40 years old, displayed excessive closeness, but with flat and uninterested affect.

Family history. No other family member could be examined. The mother had died giving birth to the twins. The father had hung himself. There were no further siblings.

Diagnoses. *M 15–1/2 according to DSM-III-R and ICD 10, catatonic type of schizophrenia.*

Leonhard Classification. *M 15–1/2 periodic catatonia.*

Subjects M 19–1 and M 19–2 (concordant C1)

M 19–1 was admitted as an inpatient upon request of his father because of alcohol abuse. He had reportedly become apathetic and lethargic and had been drinking heavily for months. For days at a time, he would lie in bed. M 19–1 exhibited sluggish and slow psychomotor behavior, and he played down his alcohol problems. In the course of his six-month hospital stay, irritability and indifference were prominent. Positive symptoms were not mentioned. *In the follow-up study,* he was 27 years old and had been unemployed for many years. His movements were slow, and he made a listless and apathetic impression. He often grimaced using his forehead. His responses were always short and vague.

The brother *M 19–2* fell ill at the age of 21. He was admitted as an inpatient as a result of his threats to commit suicide. He had locked himself in his bedroom at home, was extremely unkempt, and had been aggressive towards his parents. Upon his discharge from the hospital, he was cooperative but lacking initiative. He was readmitted because of extreme affective disturbances and further threats to commit suicide. "Peculiar movements in the shoulder area and dystonic-like movements of the torso" were observed. Depressive affect, listlessness and neglect of personal hygiene were prominent. Acoustic hallucinations and the belief that his thoughts were being heard were mentioned. Upon discharge from the hospital, he was lacking in initiative but otherwise inconspicuous. *In the follow-up study* (when he was 27 years old), he was at a rehabilitation center. Conversation with him was very laborious. Extreme blunted affect was observed. Psychomotor behavior was reduced and mild parakinesia in the right shoulder area was diagnosed.

Family history. The mother and father were personally examined. The parents were both a bit mistrustful and reserved. No symptoms of any psychological illness were found. A two years older brother was reported to be healthy.

Diagnoses. *M 19–1 according to DSM-III-R and ICD 10, residual type of schizophrenia, M 19–2 schizoaffective disorder.*

Leonhard Classification. *M 19–1/2 residuum with periodic catatonia.*

Subjects M 25–1 and M 25–2 (concordant C3)

M 25–1 developed an erotomanic delusion at the age of 27 years. After his advances had been curtailed by threat of legal punishment, he was twice treated in a general hospital with the diagnosis of "vegetative dystonia". When he was 28 years old, he did not leave the house for months at a time, and expressed occasional persecutory delusions. In the hospital, he was apathetic and lifeless. He was discharged upon improvement. At 32 years, he violently knocked down his father in an extremely agitated state. He had felt threatened by voices. He claimed to have been spoken to from the underground, and had been instructed not to reveal what had been said. He expressed various delusions; for example, that he must make furs made of hair with Mrs. J.F. Kennedy and that the furs must be finished in 10,000 years, yet after 20,000 years they were still not ready. He believed himself to be one of the few people who could make furs using hair. No positive symptoms remained upon his discharge. *In the follow-up study,* he was 57 years old and in early retirement. He had a friendly manner; however, he denied ever having been mentally ill. According to his mother, he was often tormented by voices at night which caused him to scream.

The brother M 25–2 had up to now not been in inpatient psychiatric care. He was slightly hectic but friendly. He described himself as nervous and overly resentful, and said that he lost his temper easily. He led a very secluded life with his wife and two children. He had been seeing a neurologist for several years due to chronic sleep disorders.

Family history. The 84 year old mother could be personally examined. She was physically and intellectually still in good health and showed no signs of odd behavior. The father was deceased and was said to have been healthy. A two years younger brother was reported to be healthy.

Diagnoses. *M 25–1 according to DSM-III-R and ICD 10, paranoid type of schizophrenia, M 25–2, sensitive and overly resentful personality, chronic insomnia.*

Leonhard Classification. *M 25–1 affective paraphrenia.*

Subjects W 27–1 and W 27–2 (concordant C1)

W 27–1 became ill for the first time when she was 24 years old, and was hospitalized eight times (once for a six year period) until she was 35 years old. She then worked full-time as a nurse until her retirement. Short-term inpatient stays occurred again when she was 37 and 45 years old. The phases of the illness were characterized by affective and thought disorders and "catatonic" symptoms. The diagnoses made fluctuated between hebephrenic and catatonic schizophrenia. Examples: "mute, lies on the bed, hardly moves", ... "shallow-euphoric, rather silly, confused train of thought" ... "dances and sings, coy and stilted" ... "looks helplessly about, is not capable of answering questions sensibly" ... "restlessness, foolish and inappropriate cheerfulness". *In the follow-up study* the subject was 63 years old and healthy. She displayed a warm manner and described various interests.

The sister *W 27–2* was in inpatient care twice. When she was 36, she became "catatonically" excited. She heard the voice of God and remained motionless for hours. She became healthy again after two weeks. Three years later she was readmitted in a mute and rigid state. She again heard the voice of God. A remission occurred again. She worked actively as a social worker until her retirement. *In the follow-up study the subject* was 63 and healthy, and took an active part in life, like her sister.

Family history. No family member could be examined. The mother had died at the age of 51 from cancer, and was reported to have been mentally healthy. The father, who had died when he was 78, had never been mentally ill either. A two years older sister was reported to be healthy.

Diagnoses. *W 27–1 according to DSM-III-R and ICD 10, catatonic type of schizophrenia, W 27–2 according to DSM-III-R, schizophreniform disorder, according to ICD 10, acute transient psychotic disorder.*

Leonhard Classification. *W 27–1/2 cycloid psychosis (motility psychosis).*

Subjects W 29–1 and W 29–2 (concordant C1)

The twins *W 29–1/2* became acutely ill simultaneously, with an identical psychosis, at the age of 18. Severe affective and thinking disorders were prominent, including a pronounced psychomotor hyperkinesia: "They enter the room grimacing like clowns" ... "the sisters tap their thighs, roll their eyes, hop about on their tip-toes" ... "they writhe and lash out with their hands and feet" ... "their excited state is juxtaposed with short-term states of total apathy and tension" ... "they use nonsensical and incoherent speech" ... "they make faces, grimace, and speak in a completely confused manner" ... "the twins giggle foolishly, are over-affectionate and mistake people" ... "they show irrational behavior, move awkwardly and stiffly". A full remission occurred with both subjects. *W 29–1* was in inpatient care twice more at 20 and 28 years with identical symptoms. *W 29–2* fell ill repeatedly with extreme motor activity or hyperkinesia, requiring only outpatient treatment. In addition, delusions of reference and persecutory delusions occurred. When she was 49, an inpatient stay was necessary because of religious delusions, depressive affect and psychomotor inhibition. She became healthy in a short time. *In two follow-up studies* at intervals of two years, the sisters were 47 and 49 years old. Both sisters had developed a deep, almost fanatic religiousness, with fixed beliefs in the manifestation of Maria and miracle cures. However, no delusions were evident. W 29–1 was slightly livelier than her sister. Signs of active or residual "schizophrenic" symptoms were not present. The sisters lived on disability pensions (both because of damaged discs) and kept an orderly home.

Family history. The 90 year old mother could be personally examined. She was intellectually and physically still very fit and remembered much of her pregnancy with the twins. She had never been mentally ill. The father was deceased. He had always been healthy. A two years older sister was reported to be healthy.

Diagnoses. *W 29–1/2 according to DSM-III-R and ICD 10, undifferentiated type of schizophrenia.*

Leonhard Classification. *W 29–1/2 cycloid psychosis (motility psychosis with anxiety and elements of confusion).*

Subjects W 32–1 and W 32–2 (concordant C1)

W 32–1 was the first sister to become ill at the age of 19. She would suddenly laugh at everything, and refused to eat and go to work. Upon her admission to the hospital, she appeared apathetic and lifeless and was periodically completely mute. For hours at a time she would stand in the ward with a frozen smile on her face. After six months and because she was in remission, she was able to be discharged. At 25 years, she heard commanding voices, constantly complained, refused to eat, exhibited negativistic behavior or was flatly indifferent with superficial euphoria. She was discharged upon improvement. At 29 years a relapse occurred with auditory hallucinations, extreme testiness and lack of initiative. Partial remission occurred. She has been hospitalized continually since she was 52 years old. Since then she has had periods when she heard voices, was aggressively irritable and negativistic. In between those periods, she was sluggish, apathetic and uninterested. *In the follow-up study* the 59 year old subject displayed mimetic rigidity, sluggish psychomotor behavior and extremely flat affect. No positive symptoms were present.

W 32–2 became ill for the first time at 23 years. She either sat around inactively and refused to eat, or she spoke excessively, and was excited and stimulated. In the hospital, she would lay in bed apathetically, murmuring, suddenly jump up, become aggressive, and try to fight. After a short-term improvement, she heard voices, was timid and made faces. She has been continually hospitalized from the age of 32. Affective disturbances with "frightening" grimaces, negativism, aggression and swearing voices occurred repeatedly. She then became more approachable, but lacked drive and initiative. *In the follow-up study* (at 59 years) she exhibited, as did her sister, blunted affect in a flat residual state.

Family history. A three years younger brother was personally examined in an adult home for the chronically mentally ill. According to his medical records, he suffered from chronic and relapsing schizophrenia. During examination, he exhibited severe residual symptoms with periodic catatonia. The mother had already passed away. She had always been mentally healthy. The deceased father had twice been in inpatient psychiatric care with the diagnosis of an endogenous depression. A sister of the father was reported to have been "in a psychiatric hospital at least once". Two older brothers (six and two years older) and a four years older sister were all healthy.

Diagnoses. *W 32–1/2 according to DSM-III-R and ICD 10, catatonic type of schizophrenia.*

Leonhard Classification. *W 32–1/2 periodic catatonia.*

Subjects W 36–1 and W 36–2 (concordant C1)

W 36–2 came to professional attention for the first time when she was 15 years old, following week-long nervousness, confusion and delusions of reference. Her first psychiatric hospitalization was when she was 35 years old and married. She felt persecuted and harassed, and complained of anxiety and her husband's unfaithfulness. According to her medical records, the subject had suffered repeatedly from depressive affect and agitation. She was discharged upon notable improvement. The subject's second admission occurred when she was 40. She believed the stabbing pain in her chest was caused by someone else. The subject was frequently inhibited and depressed, but also accused her husband of committing adultery and making plans to murder her. She then showed signs of logorrhea, maniacal agitation and stereotypical speech. She was discharged after improvement. Her third admission was at 44 years. She spoke profusely, was aggressive, felt persecuted, and believed that someone wanted to stab her in the abdomen. She was hostile towards her husband. She was discharged after improvement, but showed flat affect and lacked drive. *In the follow-up study* she was 54 years old. With stimulated affect, she accused her husband of deceiving her and not caring for her properly. She did not disassociate herself from her delusions of persecution and intrusion.

The sister *W 36–1* was 31 years old upon her first hospitalization. She complained of anxiety and persecutory and poisoning delusions. She was aggressive towards her husband, whom she accused of intending to murder her. In the course of her illness, she was depressed. Substantial improvement was evident upon discharge. *In the follow-up study* (at 54 years) she was very reserved and mistrustful. She did not refer to her illness at all. According to her husband, she was often agitated and would talk to herself.

Family history. The 77 year old mother could be personally examined. She appeared nervous and upset. She stated that she suffered repeatedly of "nervousness, lasting for days", but never went to see a doctor about it. The father was deceased. He was said to have been easily irritated and offended. A sister of the father was "in an awful state for a whole year and would weep constantly". A brother one year younger and another eight years older were healthy.

Diagnoses. *W 36–1/2 according to DSM-III-R and ICD 10, schizoaffective disorder.*

Leonhard Classification. *W 36–1/2 affective paraphrenia.*

Subject M 37–1 and M 37–2 (concordant C3)

M 37–1 visited a school for disabled children and was admitted to a psychiatric hospital at the age of 15, a year after beginning a weaver's apprenticeship. He appeared mentally retarded and the diagnosis of a mental retardation combined with schizophrenia was made. He believed someone was trying to poison him and claimed that his father had broken his back. Acoustic hallucinations were mentioned. He behaved peculiarly, was intrusively clinging, grimaced and moved awkwardly. The subject showed inexplicable outbursts of anger. He often lay in bed exhibiting flat affect, and then would suddenly became agitated and grimacing. He improved and could be discharged. A relapse occurred when he was 18. Thereafter he was continually hospitalized. Grossly inappropriate affect occurred repeatedly with awkward and mannered psychomotor behavior. In between those episodes, the subject showed lack of drive, but was polite and affable and adapted well to occupational therapy. *In the follow-up study* he was 59 years old. He showed extremely clinging behavior and stereotypically repeated the same sentence. His articulation was hardly intelligible. He showed awkward and angular psychomotor behavior. Grimaces in the forehead and repeated twisting of his right arm were evident, as was an extremely flat affect. No positive symptoms were apparent.

The brother *M 37–1*, a civil servant, has never been in psychiatric care. However, he showed very unusual behavior during the medical examination. His shirt hung out from his trousers and was improperly buttoned. During the conversation he was difficult to interrupt. His facial expressions were peculiar, and his movements awkward and disjointed, which became more pronounced in excitation. He made strange suggestions for his brother: His brother should pour coffee on the table and fling the cup against the wall in order to demonstrate independence; and he should put his feet on the table and rock his chair back and forth.

Family history. There are no further siblings. The 85 year old mother, intellectually still quite astute, could be examined in person. No signs of mental illness were found. The father was deceased. In the son's medical records, the abnormal and frightening behavior of the father was mentioned. A paternal female cousin apparently had suffered from a postpartum psychosis and another relative of the father's, diagnosed with schizophrenic psychosis, had died in an institution.

Diagnoses. *M 37–1 according to DSM-III-R and ICD 10, personality disorder not otherwise specified, M 37–2 according to DSM-III-R, disorganized type of schizophrenia, according to ICD 10, schizophrenia not otherwise specified.*

Leonhard Classification. *M 37–1 abnormal personality, M 37–2 periodic catatonia.*

Subjects W 38–1 and W 38–2 (concordant C1)

W 38–2 showed her first symptoms of mental illness when she was 14 years old. For weeks, sleeplessness, inner restlessness and the belief that others were sexually influencing her were prominent. She was hospitalized at the age of 23 with symptoms of extreme inner agitation and sleeplessness. W 38–2 felt physically changed and sexually influenced and would hear her father's voice demanding intercourse. She would constantly press her hands into her lap so that her genitals would not fall out. Delusions of reference, thought withdrawal and thought insertion persisted. Her affect was both reticent-depressive and also easily irritated. Upon discharge there was marked improvement, but phobias, such as of walking over bridges, still existed. Three times, at the ages of 24, 29 and 31, she was treated as an inpatient. Symptoms leading to her admissions were delusions of reference and of being controlled, with somatic symptoms of excessive sweating, heart palpitations and inner agitation. Acoustic hallucinations also occurred. She felt envied by others "because of her good looks" and unfairly treated. In interviews, she was friendly, but always slightly distant and mistrustful. *In the follow-up study* she was 33 years old and still not able to hold a job. She did not hesitate to answer general questions, but in the case of specific questions concerning her mental disorder, she behaved rather dismissive and was mistrustful. She denied ever having suffered from hallucinations and delusions of reference. She appeared generally lacking in energy and drive.

W 38–1 became mentally ill for the first time at 22 years. She heard voices, lost a lot of weight, began to neglect her personal hygiene and dropped out of university. She improved without receiving any medical care. The subject was admitted to a psychiatric clinic for the first time against her will, at the age of 25 years. She would lie in bed for days without changing her clothes, was very agitated and would "scream horribly". In the hospital, she complained of excessive anxiety and feeling inferior. The subject appeared reserved and would often just look at her interviewer, without saying a word. She was discharged upon improvement. During her second inpatient stay, she was 30 years old and pregnant. She complained again about feeling anxious, and also about the father of her child. She was very often stubborn, dismissive, easily agitated and resisted most instructions. She was discharged upon improvement. *At the time of the follow-up study*, W38–1 was 33 years old and had been unemployed for a long time. She exhibited pressured speech and was clearly irritated. She did not believe herself to be ill. She believed that others did not care for her properly and wanted to take away her child. She vigorously denied ever having suffered from acoustic hallucinations. According to her parents, W 38–1 appeared to have chronic ideas of reference.

Family history. The 67 year old mother and the 67 year old father were personally examined. The mother was in outpatient psychiatric care because of paranoid depressions. During the interview, she was mistrustful and resentful, taking most things very personally. The father appeared to be compulsive and overly concerned. He described himself to be unstable and over-anxious. A brother four years younger had excessive anxiety when he was 11 years old and was in outpatient psychiatric care for half a year because of a compulsive disorder. A one year younger sister was said to be healthy.

Diagnoses. *W 38–1 according to DSM-III-R, atypical psychosis, according to ICD 10, other nonorganic psychotic disorder, W 38–2 according to DSM-III-R and ICD 10, schizoaffective disorder.*

Leonhard Classification. *W 38–1/2 affective paraphrenia with a mild residual psychopathology.*

Subject M 42–1 and M 42–2 (concordant C1)

M 42–2 fell ill when he was 24 years old, two years before his brother did. When he was admitted to the clinic, he claimed that a Libyan or a member of the secret service was trying to find him, which was why he would hide from taxi drivers and the police. Despite these persecutory delusions, flat affect and considerable attention and concentration disorders were evident. A short time later he tried to strangle a male nurse, ate cigarettes, and appeared perplexed with an immobile and empty facial expression. He did not respond to questions, paced restlessly back and forth and performed purposeless activities. Later on, the subject was monosyllabic, bewildered, expressionless, and reported that three angels were standing around his bed. A few days later he displayed elevated affect, and believed that he was the new leader and German Chancellor and wanted to open a keg of beer. A major disturbance in the form of thought was evident in these elevated states. He had improved upon discharge, but a formal thought disorder and flat affect were still manifest. In the subsequent years, three further inpatient stays were necessary. Anxiety, dysphoric-irritable affect, persecutory delusions, acoustic hallucinations and major disturbances in the form of thought were prominent symptoms of this patient. At the *time of the follow-up study, he* was 31 years old and worked in a rehabilitation center. He was friendly, showed flat affect and a formal thought disorder. Delusions and hallucinations were absent.

The brother *M 42–1* fell ill at the age of 26, and was admitted to a psychiatric clinic four times before the follow-up study. Persecutory delusions, bizarre thoughts (the bells were ringing because his cat bit somebody), disturbances in the form of thought, nonsensical behavior (he would drink mouthwash in order to annoy his mother) and anxious, depressive affect were prominent in the acute phases of his illness. However, he also showed inappropriate, flat and unresponsive affect. At the *time of the follow-up study,* he lived with his mother and was unemployed. He did not go out of the house because he was afraid that people would talk about him. He was friendly, appeared affectively flat and showed disturbances in the form of thought.

Family history. The 54 year old mother and the 32 year old sister could be personally examined. The mother proved to be mentally healthy and had never been in psychiatric care. The sister had been in psychiatric care several times because of a schizoaffective disorder. After her personal examination, the diagnosis was made of a cataphasia according to the Leonhard Classification. The father had suddenly committed suicide at the age of 41 years, which had completely surprised everyone. According to his medical records, the father had tended to impulsive actions and alcohol abuse.

Diagnoses. *M 42–1/2 according to DSM-III-R and ICD 10, schizoaffective disorder.*

Leonhard Classification. *M 42–1/2 cataphasia.*

Monozygotic discordant pairs

Eight of the 22 monozygotic pairs were discordant, i.e. one subject was healthy in each case: M7, M11, M14, M16, W17, M28, W31, W35. The case histories of the ill subjects, including their family medical histories, are summarized below:

Subject M 7–1 and M 7–2 (discordant)

M 7–2 fell acutely ill at 22 years with delusions of reference and fear of being poisoned. After the subject's blood sample was taken by his family physician, he became extremely agitated and afraid for his life. In the hospital, he complained of being offended and mocked by others. He was also afraid that someone was trying to hurt him. He was discharged upon improvement. *In the follow-up study he was 30 years old.* According to his parents, he had not left the house for weeks, because he was afraid of people. He said that people looked at him suspiciously and said bad things about him. He appeared very anxious and weepy during the examination. He heard disparaging voices and expressed thoughts of committing suicide. He was then admitted as an inpatient and was discharged after four months in a much improved state.

M 7–1 was 30 years old during examination, married and employed in the public sector. He was healthy and had never been in psychiatric care.

Family history. The 55 year old mother, the 58 year old father and the three years younger sister were personally examined. They all showed appropriate behavior and had never been in psychiatric care.

Diagnoses. *According to DSM-III-R, schizophreniform disorder, according to ICD 10, acute transient psychotic disorder.*

Leonhard Classification. *Cycloid psychosis (anxiety psychosis).*

Subject M 11–1 and M 11–2 (discordant)

M 11–2 became acutely ill with extreme anxieties at the age of 26. He explained events in his surroundings in a delusional way, and believed that he must fight in the Gulf War. Sporadic acoustic hallucinations claimed that he would soon be captured. Compelling thoughts of committing suicide were prevalent. Following the separation from his girlfriend, he became euphoric and ecstatic for a short time. During his inpatient stay, anxious inhibition and depressed affect were prominent. A full remission occurred after five months. *In the follow-up study,* the subject was 28 years old and mentally stable and healthy.

M 11–1 was a 29 year old student at the time of the examination. No signs of any mental disorder were found.

Family history. The 56 year old mother and the 55 year old father were personally examined. They were both healthy and had never been in psychiatric care. Two older sisters (four and two years older) were mentally stable and healthy.

Diagnoses. *According to DSM-III-R, schizophreniform disorder, according to ICD 10, acute transient psychotic disorder.*

Leonhard Classification. *Cycloid psychosis (anxiety-happiness psychosis).*

Subject M 14–1 and M 14–2 (discordant)

M 14–1 was 19 years old at the time of his first inpatient treatment. A maniacal syndrome had led to his admission, with extreme psychomotor activity, elevated affect and exaggerated self-esteem. In the interview, he constantly paced back and forth. His speech was confused and pressured "with extreme loosening of associations almost to the point of incoherence". Furthermore, delusions of reference and persecution existed. During the course of his stay, he became quieter; however, he still appeared perplexed and related that he was being monitored and spied on by the police. He had "complicated dreams, but never felt anxious". He later reported that he felt as though he was in a movie. He also thought that he was being investigated by the police because an acquaintance of his worshiped a satanic cult. Over the course of several weeks, slight inhibition and lack of initiative were evident. Three months later, when his disorder was in the residual phase, he was discharged. He was readmitted again as an inpatient eight months later. He was edgy, uncooperative and displayed temporary inappropriate affect. Delusions of reference occurred with feelings of bewilderment. Then he became restless with a more elevated affect. After six months, he was stable and mentally healthy again. *In the follow-up study he was 22 years old and healthy.*

M 14–2 was 22 years old during examination and in military service. He showed no signs of any psychological disease and had never been in psychiatric care.

Family history. The 60 year old mother was the only relative personally examined. She was once treated as an inpatient, with the diagnosis of "involutional paranoid depression". The symptoms consisted of depression, apathy and psychomotor inhibition, delusions of reference and inner voices. During the examination the mother appeared slightly depressed; however, she indicated that she felt well at present. A brother of the mother was very depressed after the death of his wife and shot himself three years later. The 63 year old father was said to be healthy; however, he tended to abuse alcohol. Two older brothers (12 and eight years older) and three older sisters (11, nine and five years older) were all mentally healthy.

Diagnoses. *According to DSM-III-R and ICD 10, schizoaffective disorder.*

Leonhard Classification. *Cycloid psychosis (confusion psychosis).*

Subjects M 16–1 and M 16–2 (discordant)

M 16–2 was admitted as an inpatient for the first time at 18 years following an attempted suicide with an overdose of pills. He complained of a general lack of drive and interest. He believed that a neighbor whose wife had committed suicide wanted to kill him. He had constantly made observations which confirmed his fears and made him even more anxious. Eventually, the subject became convinced that the man had killed his wife and faked the suicide. Discharge occurred when the patient was in full remission. At the ages of 19, 20 and 35 years, further inpatient stays were necessary. Depressed affect and lack of initiative combined with anxious delusions of reference and persecution were described regularly. He heard disparaging voices and thought that someone was trying to poison him. He was discharged each time with insight into his illness and in a stable condition. *In the follow-up study* he was 36 years old, married and employed. He was friendly, responsive, calm, reflective and spoke openly about his past episodes of illness. Occasionally he would be temporarily controlled by anxieties, but was generally able to suppress them.

The twin brother *M 16–1* was married and employed at the time of the examination. He was healthy and had never been in psychiatric care.

Family history. The 69 year old mother, the 68 year old father and a one year older sister were all personally examined. The mother appeared healthy; however, she complained of loneliness. As all her seven children had left home, she was often depressed and weepy. The father was cheerful and active. He had never had any mental problems. The sister was healthy. Three older brothers (seven, five and four years older) and a seven years younger brother were all well.

Diagnoses. *According to DSM-III-R and ICD 10, schizoaffective disorder.*

Leonhard Classification. *Cycloid psychosis (anxiety psychosis).*

Subjects W 17–1 and W 17–2 (discordant)

Both sisters went to a school for disabled children and were described as being "less gifted". *W 17–2* fell ill after the marriage of her twin sister when they were 24 years old. She became easily excited and aggressive towards her parents. Anxiety and persecutory delusions developed. She heard threatening voices and thought that there was a war going on. She had the feeling "that her thoughts disappeared". In the hospital, she would shout stereotypically that she wanted to go home. She did not divulge any other information. She was discharged, but showed little improvement. She became stable while at home. At the age of 27 years, she became extremely distraught when she was at a dance, talking to herself and making a tortured anxious impression. After a short inpatient stay, she was discharged in an improved condition. At home, anxiety, withdrawal and eating binges occurred repeatedly. W 17–2 became increasingly aggressive towards her parents, but also had periods of relatively stable behavior. At the age of 37 years, she was brought to a home for the mentally handicapped by relatives. As before, anxiety attacks occurred frequently. According to an attendant at the home, the subject would then withdraw and talk and whimper to herself. Except for these states, she was very affectionate and manageable. *In the follow-up study* the subject was 40 years old. She was clearly retarded but otherwise healthy.

The sister W 17–1 was 40 years old during examination, married with three children ages 11, seven and six. She made a mentally deficient but healthy impression. She had never been in psychiatric care. The 11 year old son had average intelligence and showed appropriate behavior. The two younger sons were obviously mentally deficient and visited a school for disabled children. They did not show any other abnormal behavior at their examinations.

Family history. The 82 year old mother and a sister six years older were personally examined. Both were friendly and showed no signs of any mental disorder. The deceased father was said to have been a strict, active and always healthy man. A 11 years older brother was reported to be healthy.

Diagnoses. *According to DSM-III-R, moderate mental retardation and atypical psychosis, according to ICD 10, mild to moderate mental retardation and psychotic disorder not otherwise specified.*

Leonhard Classification. *Anxiety psychosis with multiple episodes, mental retardation.*

Subjects M 28–1 and M 28–2 (discordant)

M 28–2 fell ill at the age of 27. He felt that others were trying to destroy him at work. He believed that work colleagues would follow and observe him after leaving work, and that conversations that he had had with his family at home would be reported word for word the next day by his work colleagues. He also thought he was being monitored by the secret service. M 28–2 experienced these phenomena with increasing anxiety and therefore quit his job. In the hospital, he was extremely unnerved and perplexed. He felt threatened by other patients and nurses, and would say he didn't know what "these men" (the other patients) were doing there. Upon discharge he was fully remitted and looked forward to working again. *In the follow-up study* he was 30 years old. He spoke with insight about his mental disorder. The psychopathological results were normal.

M 28–1 was 30 years old at the time of the interview, married and employed. He was healthy and had never been in psychiatric care.

Family history. The 60 year old mother made a healthy impression during her personal examination. She had never had mental problems. The 66 year old father suffered from heart failure, but had always been mentally well. Two younger brothers (four and one year(s) younger) and a six years older sister were all said to be healthy.

Diagnoses. *According to DSM-III-R, schizophreniform disorder, according to ICD 10, acute transient psychotic disorder.*

Leonhard Classification. *Cycloid psychosis (anxiety psychosis).*

Subjects W 31–1 and W 31–2 (discordant)

W 31–2 fell acutely ill with persecutory delusions and extreme anxiety at the age of 28. She felt herself to be threatened by communists and was afraid that they would run her over with a car. In the hospital, she answered many questions inappropriately. There were indications of thought broadcasting. Days later she showed restless psychomotor behavior and spoke of being in a trance. She appeared depressed and complained of "wandering thoughts". Upon her discharge she had improved, but felt tired. At 42 years, she was admitted to the hospital for the second time. Exhaustion, dizziness, sleeplessness and extreme anxiety about being murdered were all prevalent. In the hospital, she anxiously believed that everyone was allied against her. She was depressed and her psychomotor behavior was restrained. Upon her second discharge she was again stable and sociable. *In the follow-up study* she was 47 years old, a married housewife, with one child. She was friendly and responsive and the psychopathological results were normal. She reported that she was periodically very anxious, not able to think properly, and inwardly very restless and uneasy.

The sister *W 31–1* was 47 years old during examination, employed, married with two healthy children (16 and 18 years old, one seen personally). She was healthy and had never been in psychiatric care.

Family history. The 79 year old mother was personally examined. She was healthy, and physically and intellectually still very spry. The father had died of a heart attack when he was 53 years old, without any evidence of a mental disorder. A sister two years younger, two older sisters (eight and three years older) and two older brothers (six and five years older) were all reported to be psychologically healthy.

Diagnoses. *According to DSM-III-R and ICD 10, schizoaffective disorder.*

Leonhard Classification. *Cycloid psychosis (anxiety psychosis).*

Subjects W 35–1 and W 35–2 (discordant)

W 35–1 fell ill at the age of 21 years, 4 weeks after giving birth to her first child. She did sudden "bizarre things" and spoke in a confused manner. In the middle of the night she would get out of bed, wake her husband and want to light a candle and walk though the woods. Everything appeared strange and altered to her. She would hear the voices of the nurses in the hospital where she had delivered her baby. In the psychiatric clinic, she was confused, very restless and spoke incoherently. She then became helpless, distraught and had to be fed. She lay apathetically in bed and looked around helplessly. She did not recognize where she was and was afraid that she was going to be operated on. Improvement

came slowly in the course of a month. Discharge occurred in a "considerably improved state". *In the follow-up study* the subject was 55 years old, a housewife, and married with two healthy children. She was healthy and had never been in psychiatric care again.
W 35–2 was 57 years old at the time of the interview, married with one child. There were no signs of any mental disorder and she had never been in psychiatric care.

Family history. The 78 year old mother showed no abnormal behavior during the personal examination. No psychiatric treatment had ever been given. The father was deceased and was said to have been healthy. A 10 years younger brother was reported to be healthy.

Diagnoses. *According to DSM-III-R, postpartum schizophreniform disorder, according ICD 10, postpartum psychosis or acute postpartum transient psychotic disorder.*

Leonhard Classification. *Cycloid psychosis (confusion psychosis).*

Dizygotic concordant pairs

Of the 25 dizygotic pairs, seven pairs belong to the concordance groups C1/2 (M12, W 21, M40, M41) and C3 (M6, W10, W43). Their case histories, including their family medical histories, are summarized below.

Subjects M 6–1 and M 6–2 (concordant C3)

M 6–2 was admitted to the psychiatric clinic for the first time at the age of 28 because of acute agitation. His delusional thoughts revolved around his approaching divorce date. Delusions of persecution and reference existed, and his train of thought was "muddled". He believed himself to have been infected with a venereal disease by a kiss, which also affected his brain and kidneys. His father-in-law supposedly threatened him. Many men appeared very similar to his father-in-law to him. Upon discharge he was improved but without insight into his illness. The second inpatient stay occurred at 43 years. In the meantime, M 6–2 had gotten divorced for the second time and had been unemployed for seven years. He was restless, tense and driven. He believed he was being irradiated and that the whole house was full of rays and vapors. He had to call the police, and everything smelled of cat excrement to him. He claimed to have been irradiated before, and that a ball of lightning had once rolled over him. Upon discharge, he still complained of different physical pains, but he appeared stable otherwise. His third inpatient treatment occurred at the age of 47. He was in retirement. He was agitated, expressed delusions of reference, indicated that he could not stand the irradiation any longer and that he could feel "surging emotions in his brain". Thought insertion and thought withdrawal existed. A noticeable lack of drive was still evident upon discharge. At the time of the *follow-up study,* the subject was 51 years old. He lived socially isolated and withdrawn, was taciturn and only hinted at his disease. He was considering moving since his apartment supposedly had many things wrong with it. He evaded the question as to whether he was still being irradiated.

The M 6–1 brother was 52 years old, married and employed at the time of the examination. He seemed nervous and fidgety, and described himself as sensitive, unstable and unbalanced. He had been seeing a psychotherapist for years.

Family history. The 76 year old mother was personally examined. She was friendly and gave information readily. She was often "a nervous wreck". She had never been in psy-

chiatric care. The father was killed in World War II. A brother of the paternal grandfather had committed suicide. A two years younger sister was healthy.

Diagnoses. *M 6–1 according to DSM-III-R, avoidant personality disorder, according
to ICD 10, anxious-avoiding personality, M 6–2 according to DSM-III-R and ICD 10,
paranoid type of schizophrenia.*

Leonhard Classification. *M 6–1 emotionally unstable-anxious personality, M 6–2 affective paraphrenia.*

Subjects W 10–1 and W 10–2 (concordant C3)

W 10–1 was admitted for the first time as an inpatient after an attempted suicide with
tablets at the age of 18. She had fallen in love with a man who apparently did not want a
relationship with her. She did not describe her motives very coherently, so that a differen-
tial diagnosis of a psychosis was also made. Nothing further was recorded in her files.
Inpatient treatment did not take place again. *In the follow-up study* she was 48 years old
and lived reclusively in a small apartment. She was divorced and had been unemployed
for several years. She was friendly and responsive in the interview, and reported that she
intended to marry. She had not seen her intended husband for a long time, but he gave her
secret signs in newspaper advertisements. Other psychopathological phenomena could
not be explored. Months later she complained several times by telephone that she had
been deceived. She did not give more precise information.

   W 10–2 had never been in psychiatric care. At the interview, she was at first very reserved. During the course of the conversation, she became more communicative. She
described herself as sensitive and touchy, and that she was often unbalanced and very
depressed for weeks on end. She could then not tolerate being with anyone and would
withdraw. Then she began in the interview to talk increasingly volubly, and spoke about
her philosophy of life and ideals, which contained many references to metaphysical and
magical elements. Occasionally, near-delusional ideas were observed.

Family history. According to W 10–2, the now 79 year old mother suffered from long
depressive phases, during which she secluded herself. At other times, the mother dressed
and made herself up in a rather bizarre manner and was almost "fanatically hyperactive".
During the personal examination, the mother was a bit condescending but amiable. She
had never been in psychiatric care. She denied having any mental problems. The father
was deceased. He had often been irritable and aggressive. He would tyrannize the family
and always wanted to move in the best circles. A three years older brother reportedly had
alcohol problems. A one year younger sister was said to be "insanely religious, bigoted,
narrow-minded and egocentric". The father of the mother had hung himself.

Diagnoses. *W 10–1 according to DSM-III-R and ICD 10, delusional disorder, W 10–2
according DSM-III-R, borderline disorder, according to ICD 10, emotionally unstable
personality.*

Leonhard Classification. *W 10–1 strong suggestion of a mild form of affective paraphrenia, W 10–2 abnormal personality.*

Subjects M 12–1 and M 12–2 (concordant C1)

The twins were two and a half years old when the mother died. They grew up in an
orphanage until they were 15 years old, both achieving a middle school certificate.
After this, they lived in various dormitories for apprentices, repeatedly leaving their just-

begun apprenticeships. *M 12–1* came into inpatient psychiatric care for the first time at the age of 17 because of compelling thoughts of committing suicide. He lacked drive, was dysphoric, and showed reduced facial expressions and gestures. He was very taciturn during conversations. After four months, discharge was possible because of improvement. A second admission occurred at 18 years. He said he heard a masculine voice issuing commands, such as to smash vases. Upon admission, he again lacked drive, was dysphoric and reacted with negativism. During the course of his stay, psychomotor excitement with aggression occurred repeatedly. Now and then, he was withdrawn and reclusive; however, he took part in occupational therapy. After 15 months, he was discharged to a rehabilitation center in an improved state. Two further inpatient stays followed because of aggressive excitement. *At the time of the follow-up study,* he was 22 years old and living in a group home for the mentally ill. He was dysphoric but not unfriendly. In the interview, he spoke very little and lacked drive. He appeared apathetic, lethargic, with blunted affect. He grimaced constantly in the eye and forehead area of his face. No positive symptoms were present.

No inpatient treatment of the brother *M 12–2* was known. He had been living in a center for occupationally rehabilitated juveniles for several years. According to a social worker, he often showed "abnormal aggressive behavior" with "conspicuous twisted movements". Otherwise, he showed appropriate behavior, but had few interests. *During the personal examination,* he was 22 years old. Decreased psychomotor activity and indifference were prominent. His facial expression was rigid. Positive symptoms were absent. Shortly after the interview he suddenly disappeared from the center. On subsequent inquiry, he was still missing.

Family history. The 60 year old father and a six years older sister were personally examined. The father, at the age of 49, had been in inpatient psychiatric care. He had felt threatened, heard voices, and was unkempt. Alcohol abuse was assumed; however, the patient, denied this. He was discharged upon improvement. At the time of the personal examination, he lived with his daughter. He had been in retirement for several years. He was taciturn, and knew nothing about the illness and circumstances of his sons. He seemed to be unconcerned about them. He was unkempt and gave the impression of being in a residual phase of a schizophrenic psychosis. The married sister of the twins appeared normal during the interview. Two other older sisters (eight and seven years older) were married and healthy.

Diagnoses. *M 12–1/2 according to DSM-III-R and ICD 10, residual type of schizophrenia.*

Leonhard Classification. *M 12–1/2 residual state with periodic catatonia.*

Subjects W 21–1 and W 21–2 (concordant C1)

W 21–1 was 48 years old when she was brought to the emergency room of a psychiatric hospital after an attempted suicide by taking tranquilizers. She expressed great anxiety that her husband wanted to shoot her. The day before she was admitted, she had jumped from her balcony and shouted, "You all want to burn me". She told her family doctor that she had lung disease and cancer. On the ward she cowered in a corner and was afraid that rifles were being aimed at her. During the course of her stay, she was clearly depressed. She also felt that she was electrified, and heard "how the nurses talked about her arrest". Discharge occurred when the patient was in full remission. A second inpatient stay occurred at 56 years. She had become sleepless and was wandering around town in a delusional state, expressing delusions of persecution and meaning. Her clothes had supposedly been arranged in a specific order and her stockings had suddenly fallen to the floor.

She said that she now knew the truth. Discharge occurred when the patient was in full remission. Further inpatient stays occurred at 58, 60, 61 and 62 years. Anxiety and delusions of persecution and reference were prominent. Accusing phonemes occurred. Affect was occasionally severely depressed. The phases of her disorder tapered off each time without residue. *During the personal examination,* W 21–1 was 66 years old and widowed. She was friendly and responsive and spoke of her illness with insight.

W 21–2 was 52 years old when she first became ill. She had also attempted suicide. In the hospital, she complained of anxiety, inner restlessness and feeling miserable. She was stable upon discharge. Two weeks later restlessness, sleeplessness, extreme anxiety and oppressive feelings arose. Full remission then occurred. A third inpatient stay occurred at 56 years because of a similar episode. At the age of 61 years, she became afraid that her relatives wanted to kill her. Delusions of reference and becoming poisoned and menacing phonemes occurred. *During the personal examination,* W21–2 was 66 years old and in inpatient care again because of similar symptoms. She could once again be discharged in remittence.

Family history. A first-born triplet sister had died at the age of 48 years of cancer. She had always been mentally healthy. The mother had died while giving birth to the triplets, and had never been in psychiatric care. The father, who had re-married, was also deceased. He had had no mental problems. A four years older brother was reported to be healthy. Three younger half-sisters were also healthy.

Diagnoses. *W 21–1/2 according to DSM-III-R and ICD 10, schizoaffective disorder.*

Leonhard Classification. *W 21–1/2 cycloid psychosis (anxiety psychosis).*

Subjects M 40–1 and M 40–2 (concordant C1)

M 40–1 was admitted as an inpatient for the first time at the age of 17 years because of depressive affect. He became well again and was discharged, and later successfully passed his exams as a retail assistant. Three years later he became dejected again, was distressed and spoke very little. At the age of 23, he was readmitted as an inpatient. He was anxious, shy, depressed and showed a severe inhibition of thought. He was discharged upon improvement. One year later, he was admitted with elevated affect and was incoherent. During the course of his stay, his behavior alternated between being infectiously engaging with nearly incoherent speech and being in an inhibited and perplexed state. He was discharged upon noticeable improvement. In subsequent years, many inpatient stays occurred. The subject could always be discharged in a healthy, or at least very much improved state. The symptoms he showed in his inpatient stays were variform. Central to his psychopathology was a consistently severe disturbance in the form of thought. He often showed a pressured speech with incoherent and incomprehensible content. On the other hand, he would also appear extremely inhibited, perplexed and mute. Either euphorically elevated or depressively anxious affect was constantly present. Acoustic hallucinations in the form of threatening and terrifying voices, brief periods when he would mistake people and extreme psychomotor activity occurred. *In the follow-up study* he was 60 years old. He lived with his sister and had helped his older brother on his farm for many years. He had lost much weight, had an anxious and perplexed facial expression and spoke very quietly and hardly intelligibly. His condition was one of inhibited confusion, which, according to his sister, had lasted for many weeks.

M 40–2 was 21 years old during his first inpatient treatment. He was irritated, talked continually and showed inappropriate behavior; for example, suddenly taking his shoes and socks off in church. During his stay, a systematic investigation was hardly possible;

he showed either elevated or irritated affect. He was in full remission upon discharge. Two years later he was admitted again with elevated affect and psychomotor agitation. Several days later, he lay in bed with a blank facial expression. Again, full remission occurred. One year later he was again in the clinic with elevated affect, irritated, agitated, speaking continuously and incoherently and showing extreme psychomotor activity. Full remission then occurred. At 34 years, he was admitted as an inpatient for the fourth time. Mood swings, pressured speech with formal thought disorder and psychomotor unrest characterized the course of the illness. A full remission occurred again. *In the follow-up study* he was 60 years old, retired and lived with his wife in their own home. He was mentally healthy and had not been in psychiatric care since his last discharge.

Family history. The parents of the twins were deceased at the time of the follow-up study. They had always been mentally healthy. The four years older brother and the eight years older sister were examined in person. The sister fell ill at 22 years with a depressive psychosis, which, according to the family's description indicates an inhibited confusion psychosis. In the follow-up study, the sister showed signs of a demential illness. She had not been in psychiatric care again. The brother fell ill with a depression at the age of 64 years, with inhibition of thought the most prominent symptom.

Diagnoses. *M 40–1/2 according to DSM-III-R and ICD 10, schizoaffective psychosis.*

Leonhard Classification. *M 40–1/2 excited-inhibited confusion psychosis.*

Subjects M 41–1 and M 41–2 (concordant C1/C2)

M 41–1 came to professional attention for the first time when he was 27 years old. He felt that he had a calling to be a psychic healer and that he could heal others by the touch of his hands. Once he was under medication, the delusions abated. Further relapses followed. He was admitted at 29 years in an extremely agitated state and had heated debates with the admitting physician. Upon discharge, rambling speech and ideas of meaning were still apparent. When he was 32 years old, his illness became chronic. Grandiose and perse-cutory delusions appeared, which continued and expanded. He saw himself as a great artist who was being challenged by different powerful organizations, such as a music and drug mafia. He believed himself to be a guardian of the law and a "Mr. Clean" and kept reporting people to the police. He suspected friends of his of terrorism or accused them of earning additional income by pornography. The subject always reacted extremely agitated when criticized or reproached, and threatened to kill his parents, brother and physicians by shooting them. He sent absurd job applications to movie stars and poli-ticians. *In the follow-up study* he was 36 years old and lived in a very neglected condi-tion in state housing. He maintained his pathological ideas and was agitated and para-noid. Due to his disturbing the peace, an attempt was being made to commit him to a psychiatric hospital.

M 41–2 was admitted as an inpatient at the age of 20 because of a paranoid-hallucinatory psychosis. He was reserved and mistrustful during admission, grimacing and exhibiting a formal thought disorder, thought broadcasting, thought insertion and thought withdra-wal. Acoustic hallucinations occurred. He was difficult to motivate, and after a short time vehemently insisted on his discharge, which occurred upon some improvement. *In the follow-up study* he was working and married. He appeared reserved, mistrustful and answered questions briefly but adequately. A disturbance in the content or form of thought did not exist; however, it was conspicuous that he believed his brother to be mentally healthy and that his brother's partially bizarre delusions were an understandable reaction to his difficult social environment.

Family history. The 73 year old father and the 71 year old mother could be personally examined. The father appeared to be resentful, irritable, dogmatic and choleric. The mother was very quiet and reserved and had been treated twice as an outpatient, because she had been depressed and had cried a great deal. Her symptoms indicated an endogenous depression. A two years older sister was working, mentally stable and had never been in psychiatric care.

Diagnoses. *M 41–1 according to DSM-III-R/ICD 10, paranoid type of schizophrenia, M 42–2 according to DSM-III-R, schizophreniform disorder, according to ICD 10 acute transient psychotic disorder.*

Leonhard Classification. *M 41–1 affective paraphrenia, M 41–2 mild residual state with unsystematic schizophrenia.*

Subjects W 43–1 and W 43–2 (concordant C3)

W 43–1 was hospitalized for the first time at the age of 20 because of a "drug induced psychosis with intellectual impairment". A second hospitalization with a different diagnosis of "hebephrenic schizophrenia" occurred at 21 years. The medical records of both inpatient stays were not informative. The third inpatient admittance was necessary because of threatened violence to herself and others. She had been discovered rampaging in her apartment. In the hospital she shouted that they had taken away her child, that everything was controlled by the mafia, and that her television had been programmed wrong. She was very agitated, aggressive and threatening. Two days later she was lethargic, hardly cooperative and gave only monosyllabic answers. At the time of her discharge she had improved, but was brought back the next day. Her social worker had found the subject in a "devastated condition", "totally rigid with wide open eyes" in her apartment. In the hospital she was clinging and showed peculiar and irritated affect. Discharge occurred after improvement. When she was 24, she was hospitalized three times, due to extreme psychomotor activity, affective disorders (peculiar, irritated, inadequate and rejecting) and disturbances in the form of thought, especially involving loosening of associations. During her inpatient stays, extreme psychomotor activity alternated with a lack of drive.

She had always improved upon discharge, but with noticeable flat affect. *In the follow-up study* the subject was 26 years old. An extreme mental impairment with flat affect and reduced psychomotor activity were evident. She denied any positive symptoms.

W 43–2 had never been in inpatient psychiatric care, but had been under supervision for years. *At the time of the follow-up study*, she lived in a supervised group home and worked in a rehabilitation center. She appeared indolent, with blunted affect and showed no interest in her twin sister or other family members. Her psychomotor behavior was awkward and lethargic. Positive psychotic symptoms were not evident. According to the social worker, she tended towards impulsive actions, aggressive outbursts and would lay in bed for days.

Family history. The 69 year old mother, three older brothers (seven, nine and 11 years older) and two older sisters (12 and 14 years older) were examined in person. The mother was in a nursing home and showed affective blunting with a suspected periodic catatonia. Two older brothers lived in supervised homes and also suffered from periodic catatonia. The other siblings were healthy. The father had passed away and supposedly had had alcohol problems.

Diagnoses. *W 43–1 according DSM-III-R/ICD 10, residual type of schizophrenia, W 43–2 personality disorder not otherwise specified.*

Leonhard Classification. *W 43–1 periodic catatonia, W 43–2 abnormal personality (residual state with periodic catatonia).*

Dizygotic discordant pairs

Of the 25 dizygotic pairs, eighteen pairs were discordant: M3, W4, M5, M13, M18, W20, W22, W23, W24, W26, M 30, W33, M34, M39, W44, W45, M46, M47. Their case histories, including their family medical histories, are summarized below:

Subjects M 3–1 and M 3–2 (discordant)

M 3–2 had always lagged behind his brother in his infant and childhood development. His brother had graduated from high school, and he had had to repeat a class in elementary school. When he was 14, his parents had started to notice that he was different from other children, that he would withdraw and hardly make any contact with others. He became "peculiar, aggressive, very restless, always moving, constantly murmuring to himself, suddenly not speaking and giving only brief answers to questions". He was put in a home for the disabled and was brought to a psychiatric hospital for the first time at the age of 23 because of aggressive agitation. In the hospital, he appeared anxious, paced back and forth and "was unable to be focused in conversation". He was discharged without much improvement. *In the follow-up study* he was 40 years old and had been permanently hospitalized for many years. He paced restlessly back and forth, constantly touching objects, walls and doors, and when addressed, would only glance up briefly. He mumbled continually to himself and seemed to be influenced by hallucinations, as he would look around, appearing to listen, and repeatedly exhibiting different facial expressions. He did not answer questions; instead, he mumbled incomprehensible words more loudly. An incessant involuntary grasping was evident in his psychomotor examination (he gave the examining physician repeatedly his hand, even when firmly requested not to do so). According to the nursing staff, his behavior had not changed since his admission, and he would become agitated at regular intervals that lasted only for a short time.

M 3–1 had never been in psychiatric care. He was 39 years old, married with two children, and working.

Family history. The 70 year old mother, the 69 year old father and two younger brothers (nine and four years younger) were personally examined. None showed abnormal behavior or had been in psychiatric care. Three younger sisters (six and three (twins) years younger) and an eight years older brother were all healthy.

Diagnoses. *According to DSM-III-R, undifferentiated type of schizophrenia, according to ICD 10, schizophrenia not otherwise specified.*

Leonhard Classification. *Systematic catatonia with sluggish and proskinetic components.*

Subjects W 4–1 and W 4–2 (discordant)

W 4–1 was admitted as an inpatient when she was 18 years old after an attempted suicide with tablets. She was newly married and pregnant. Although her husband looked forward to having the child, she blamed herself for getting pregnant. When asked questions, she turned her face away and answered "hesitantly and fragmentally". The rest of her medical records were scanty. Her diagnosis upon discharge was "reactive affect with a primarily psychopathological personality". Further inpatient stays were unknown. *In the follow-up study* she was 43 years old, married with one child. She appeared nervous, hectic but also friendly. However, when asked about her hospital stays, she became agitated. She implied that her husband was behind it all, and that he constantly sexually harassed her. Further, she stated that he still tried to harass her telepathically. She became agitated and did not want to talk about the subject. She could, however, discuss other topics quite well.

At the interview *W 4–2* was 43 years old, a housewife, married with four children. She was healthy and had never been in psychiatric care. She had little contact with her twin sister, who was often nervous and neurotic.

Family history. The 73 year old mother was personally examined. The interview was difficult. She was mistrustful and reserved. She stated that she had had a difficult and stressful life. She evaded questions about psychological problems. The father had been deceased for 25 years. He had had an alcohol problem and was very aggressive towards his family. A nine years younger sister and a four years older brother were healthy.

Diagnoses. *According to DSM-III-R and ICD 10, delusional disorder.*

Leonhard Classification. *Affective paraphrenia in a paranoid phase.*

Subjects M 5–1 and M 5–2 (discordant)

M 5–1 attracted the attention of his parents because of his abnormal behavior when he was around 22 years old. He became conspicuously withdrawn, reclusive and often so extremely aggressive that his whole family was afraid of him. He would often only leave his room to eat. He was dismissed from the army because of an "eye and nervous complaint". His first inpatient stay was when he was 28 years old. His room at home was a wreck. Rubbish and pieces of paper with very small handwriting on them lay about. When he was asked about this, he said that he was dealing with the problems of hog-raising. In the hospital, he lacked insight into his illness, and constantly complained and bickered. He complained stereotypically of eye defects, headaches and an inner restlessness. He continually asked when he would be discharged. When he was given attention, he often showed a "wide smile". He was discharged with no improvement. A second stay in the same year took a similar course. He lay apathetically in bed and complained about physical symptoms that could not be medically confirmed. At the age of 35 he was admitted to the clinic for the third time. In the meantime he continued to live at home with his parents and was unemployed. He sometimes treated his parents quite awful for example cutting his father's pigs' ears half off. *During this stay, he was examined within the framework of the study.* Stereotypical complaints and protests with no affective correlation were prominent. He constantly asked to be discharged and complained incessantly of physical symptoms despite medical reassurance. Occasionally he showed a psychologically un-motivated and somewhat peculiar smile.

M 5–2 was 36 years old during examination, married with four children, and working. He was healthy and had never been in psychiatric care. The 66 year old mother and the 66 year old father were healthy during the personal examination. There were no further siblings.

Diagnoses. *According to DSM-III-R, disorganized type of schizophrenia, according to ICD 10, Schizophrenia not otherwise specified.*

Leonhard Classification. *Systematic hebephrenia (eccentric-silly).*

Subjects M 13–1 and M 13–2 (discordant)

M 13–1 had had a normal childhood development up to the age of two. Until then, there was absolutely no difference between the twins. Then, due to a dislocation of the hip, the subject had to be treated at an orthopedic clinic. After he was discharged, he had to lie in a plaster bed for a further six weeks. According to his mother, this twin was different since that time and in particular, made no progress in speaking. He increasingly lost contact with his environment, became autistic and often showed short phases of aggressive destructiveness. In contrast, his physical development was normal. At the age of five the subject had a thorough neurological examination with normal results, and later tests (at 40 years: normal neurostatus; no indication of dysmorphia, chromosomal anomaly, and fragile-X-syndrome; normal results of blood, liquor, EEG, and CT,) all showed no indication of any organic illness. Due to increasing behavioral problems, he was admitted to a psychiatric clinic at the age of six. He was completely mute, and showed extreme motor unrest. He was always alone, fiddling with toys or paper cartons "but not in a way that one could describe as meaningful play". Severe auto-aggression and minute-long screaming attacks developed. At the age of 13, he was described as very agitated, irritable, "inexplicably" extremely excitable and aggressive, pulling faces and grimacing, running stereotypically down the halls, masturbating impulsively and shamelessly. At 15 he "still made no attempt to speak". When he was 18, he was described as "practically isolated, sits at his place rigidly, without moving nor interest". He was then placed in a nursing home without improvement. The following behavior was described there: "He straightens out everything that he sees; no initiative; outwardly appears not mentally deficient; has fixed habits; he can stand for hours in one place without moving; he has never spoken a word; becomes totally exhausted when walking in the garden, because he tries to straighten out all the leaves that lie about". *In the follow-up study* the 40 year old subject maintained a rigid posture with an expressionless face that was turned away from the interviewer. When addressed, he turned briefly towards the interviewer, but avoided eye contact and turned away immediately. He was completely mute. He touched everything he saw and moved it slightly to another position. When the interviewer tried to make eye contact with him, he turned away. When a hand was held out to him, the subject would impulsively grasp at it, even when told to stop (this symptom is know in German as 'Gegengreifen'). Every slight pressure against his body would result in excited motor movements until the pressure stopped. This resulted in the subject assuming many awkward positions. When the pressure stopped, he returned to his original position (this impulsive response is know in German as 'Mitgehen'). When not occupied, he would remain motionless.

M 13–2 was 42 years old during the personal examination. He was divorced, had been living with a partner for many years, had one child, and was employed. He was open, friendly, and with no signs of any psychological illness, and had never been in psychiatric care.

Family history. The 72 year old mother and a four years older sister were personally examined. They were healthy and had never been in psychiatric care. A six years older brother was reported to be healthy. The father was deceased. He was said to have always been a very caring father.

Diagnoses. *According to DSM-III-R and ICD 10, the diagnosis of an autistic disorder from the category of pervasive developmental disorders suggests itself. However, also according to DSM-III-R and ICD 10, unmistakable catatonic symptoms (mutism, negativism, stereotypical posturing, purposeless non-directional psychomotor excitement) justify the diagnosis of a catatonic type of schizophrenia beginning in early childhood.*

Leonhard Classification. *Characteristic symptoms of a systematic catatonia beginning in early childhood were evident, especially negativistic, proskinetic and manneristic components.*

Subjects M 18–1 and M 18–2 (discordant)

M18–2 became ill at the age of 17 with a "paranoid-hallucinatory psychosis". He had suddenly become aggressive and "behaved like a Kung-Fu fighter". He heard "evil voices" that commanded him to do bad things. Sometimes the voices said something pleasant, such as that he should marry. He believed he could influence the rolling of billiard balls and was convinced he could hypnotize animals at the zoo. Most prominent, however, was his psychotic anxiety: "he walks with sheer terror towards the ward exit and bangs against the door...appears extremely frightened, his eyes are wide open with fear". Two months after he was admitted, it was possible to discharge him in a clearly improved state. *In the follow-up study* he was 25 years old, was planning to get married, and was employed. He was friendly, responsive, gave information readily, and spoke with insight about his period of illness.

M 18–1 was 24 years old at the time of the interview, single and a student. He made a mentally stable and healthy impression and had never been in psychiatric care.

Family history. The 54 year old mother, the 56 year old father, a three years older sister and a three years younger sister were personally examined. The mother was temperamental and slightly hyperthymic, the father was rather shy and introverted. Neither appeared mentally ill. The examined siblings were also healthy. Two other older sisters (eight and three years older) and a six years younger brother were also said to be healthy.

Diagnoses. *According to DSM-III-R schizophreniform disorder, according to ICD 10 acute transient psychotic disorder.*

Leonhard Classification. *Cycloid psychosis (predominantly anxiety psychosis).*

Subjects W 20–1 and W 20–2 (discordant)

W 20–2 was 24 years old when she received psychiatric care for the first time. She had been unemployed for a long time. Upon admission, she appeared somewhat anxious and described "in longwinded, carefully chosen and bizarre wording" unspecific pain in her jaw. She believed that the pain was influenced by a minister, who was against her. Stubbornly and crossly with "slightly depressive affect", she complained of physical pains despite medical reassurance, while her ideas of being controlled became less prominent. She was then discharged without improvement and a short time later readmitted. Monotonously, she complained again of physical pains, now in the area of the back of her tongue. She was cantankerous with depressed affect and talked non-stop about her physical pains. Without great improvement, she was discharged to a rehabilitation clinic. She was again admitted as an inpatient when she was 34 years old and *was examined within the framework of the study.* Again, medically not confirmed physical pains were expressed, which the subject complained of in a slightly depressed and monotonous manner. It was not possible to discuss other topics with her. She stated that since her last discharge her complaints had never really gone away. She exhibited flat affect, and had no interests and plans for the future. There was no evidence of psychomotor abnormalities. No paranoid-hallucinatory symptoms were apparent. Upon discharge, the physical pains were less prominent, but the overall clinical picture remained the same.

W 20–1 was 34 years old, single and employed at the time of examination. She had never been in psychiatric care.

Family history. The 59 year old mother was personally examined. She talked a lot and it was difficult for her to focus on specific questions. Apart from this, she appeared normal and had never been in psychiatric care. The 63 year old father was said to be pathologically jealous. For this reason, the mother had gotten a divorce. A three years older brother was said to be stable and mentally healthy.

Diagnoses. *According to DSM-III-R, disorganized type of schizophrenia, according to ICD 10, schizophrenia not otherwise specified.*

Leonhard Classification. *Systematic hebephrenia (eccentric).*

Subjects W 22–1 and W 22–2 (discordant)

W 22–1 was treated as an outpatient for the first time when she was 15. She had been living as an orphan for two years (see family history), in an orphanage and came to professional attention because of her extreme shyness and social isolation. During the examination, she was mute and never looked at the examining doctor. At the age of 18, she was admitted as an inpatient because of an attempted suicide by cutting her wrists. During her stay, she was negativistic and gave very sparse answers, keeping her head turned away. Upon discharge, she was more amiable but still lacked drive. During her second inpatient stay, she was 21 years old and was examined *within the framework of the study.* She lay mute in bed and rejected any attention. She often grimaced using her forehead. According to her medical reports, she had repeatedly expressed anxious delusions of reference. No anxiety was detectable during this examination. She showed a rather rigid facial expression. During the course of her stay, she became more responsive and was discharged in an improved state.

W 22–2 was 24 years old during examination, had been living with a partner for several years and had two children. She appeared healthy and had never been in psychiatric care.

Family history. Other family members could not be examined. The mother died when she was 40 years old from bronchial cancer. She had always been mentally healthy. The father had hung himself in the course of a chronic paranoid-depressive psychosis. He had been retired for many years at the time of the suicide because of his mental illness and had been repeatedly in inpatient care. His symptoms included extreme inhibition to the point of stupor, anxiety and ideas of having sinned. Often, only a partial remission occurred. The acute phases of the disorder were usually accompanied by a general lack of energy and drive. A three years older brother of the twins often had anxiety attacks and had been because of this in outpatient psychiatric care several times.

Diagnoses. *According to DSM-III-R atypical psychosis, according to ICD 10, psychotic disorder not otherwise specified.*

Leonhard Classification. *Intermittent periodic catatonia.*

Subjects W 23–1 and W 23–2 (discordant)

W 23–1 was 26 years old when she was admitted to a psychiatric clinic for the first time. She reported a "nervous breakdown". Everyone was watching her, she could not sleep anymore and shook incessantly. She wanted to kill herself by throwing her hairdryer in the sink. She appeared depressed and showed reduced facial expressions and gesticulations. In addition, she lacked drive and was apathetic. She complained of anxiety and ideas of reference. She was discharged upon improvement. At 27 years, she came to the clinic for the second time. She was almost mute, and testified to brooding compulsions and anxieties of being a failure. During the course of her stay, strong anxieties and ideas of reference occurred repeatedly. She was stable upon discharge. *At the time of the follow-up study* the subject was 30 years old. She was quiet, appearing introverted, friendly and cooperative. She felt well and was mentally stable and healthy.

W 23–2 was 31 years old when examined. She lived in a stable relationship and was employed. She was very lively, friendly and showed no inappropriate behavior.

Family history. The 56 year old mother, the 61 year old father and a five years older brother were all personally examined. The mother made a slightly undifferentiated impression, and her face appeared puffy. In the daughter's medical files, the suspicion of the mother's alcohol abuse had already been recorded. This suspicion was now confirmed. However, she denied any psychological problems and had never been in psychiatric care. The father and the brother showed normal behavior. A five years older sister (twin of the brother) was said to be healthy.

Diagnoses. *According to DSM-III-R and ICD 10, schizoaffective disorder.*

Leonhard Classification. *Cycloid psychosis (anxiety psychosis).*

Subjects W 24–1 and W 24–2 (discordant)

W 24–2 was admitted to a psychiatric clinic at the age of 27 because of a "paranoid-hallucinatory condition". On admission, she was tense, mistrustful, and denied having had paranoid-hallucinatory experiences. She complained of feeling tired during the day and of derealization phenomena. She stated that four weeks ago she had frequently felt happy, but in between those times had felt very anxious. She was discharged one month later when she was fully recovered. A year later she was admitted for the second time. She walked around naked and aimlessly, and was afraid that someone was trying to poison her with tablets. She said to her husband "look me in the eyes, I am God". During her examination she was very lively and responsive. She felt the power of God in her, believed that everyone loved each other and that there would never be another war. She was healthy upon discharge, but shortly thereafter was readmitted. She had expressed that she was God or an archangel, and had put her hand on a hotplate to prove this. In the hospital, she either showed elevated affect, or was depressed, anxious and expressed ideas of reference and poverty. Full remission then occurred. She became ill again at the age of 36. This time she expressed religious ideas and said that the pope appeared before her. In the hospital, she was inhibited-depressive, anxious, and believed war and the end of the world were coming. Full remission then occurred again. At the age of 42, (as a housewife, married with two children), she became ill again and was examined *within the framework of the study*. She was very anxious, weepy, and felt pursued and threatened, especially by a man from the neighborhood. She also heard threatening voices. When she was treated with understanding, her expression would always lighten up.

W 24–1 was 42 years old at the time of the examination, a housewife, married with 2 children. In the interview, she was friendly, mentally normal and asked with concern if she could get the same illness as her sister. She had never been in psychiatric care.

Family history. The 70 year old mother was personally examined. She was rather reserved, answered many questions evasively, but spoke a lot about her difficult life during the war. Psychiatric treatment had never occurred. The father was described as stubborn and that he would sometimes show sudden outbursts of anger. He supposedly had had alcohol problems for many years. An eleven years younger sister of the twins had been in outpatient psychiatric care because of frequent states of anxiety. Two other younger sisters (nine and one year(s) younger) and two older brothers (seven and two years older) were healthy.

Diagnoses. *According to DSM-III-R and ICD 10, schizoaffective disorder.*

Leonhard Classification. *Cycloid psychosis (anxiety-happiness psychosis).*

Subjects W 26–1/2 and W 26–3 (discordant)

W 26–3 became ill when she was 18 years old, when she was training to be a nurse's aide. She was brought to the hospital by her head nurse, was "extremely confused" and did not know what she was doing or where she was. Disturbance in the form of thought including "conspicuous incoherent speech" with an at first "peculiar" and then an uncertain and perplexed facial expression were prominent during the examination. She believed sounds were coming from the wall and felt that they penetrated her brain through the back of her head. The diagnosis of a hebephrenia was made. When discharged, she was considered to be "well into remittence". One year later she was in a serious car accident and suffered from polytrauma, basal skull fracture including a complicated meningoencephalitis and cerebral abscess. After undergoing neurosurgery she got better. According to her mother, her behavior was the same as before the accident. One year later she became suddenly perplexed again, was unstable, and exhibited restless agitation with purposeless movements. "Considerable speech and orientation disorders" were present. After two weeks, she suddenly became fully orientated again and reported visual and acoustic hallucinations. Two weeks later she was discharged in a stable condition. *In the follow-up study* she was 33 years old and retired as a result of her physical injuries from the accident. She was friendly and responsive, and displayed normal content and form of thought. The two other triplet sisters *W 26–1/2* were 33 years old and both mentally stable and healthy during the examination and had never been in psychiatric care. W 26–1 was single, working, and living at home. W 26–2 was married with two children and worked part-time.

Family history. The 56 year old mother and a six years younger brother were personally examined. The mother was friendly and showed appropriate behavior. According to the medical records of her daughter, the mother had suffered from depression immediately after the birth of her first child. She revealed this information, however, only at her second interview. The younger brother of the triplets was healthy. The father had died of cancer at the age of 53 years. He was said to have been a hardworking and capable man. The father of the mother had hung himself "as a reaction to extreme financial problems". A two years older brother of the triplets was said to be healthy.

Diagnoses. *According to DSM-III-R atypical psychosis, according to ICD 10, acute transient psychotic disorder.*

Leonhard Classification. *Cycloid psychosis (confusion psychosis).*

Subjects M 30–1 and M 30–2 (discordant)

M 30–2 was admitted as an inpatient for the first time when he was 17 because he considered committing suicide. He made a dysphoric, slightly depressed impression and complained of thought broadcasting, depersonalization and derealization phenomena. Everything seemed unreal and strange to him. Everyone supposedly had pity with him because of his "terrible condition", especially his "repulsive nose". During the course of his stay, he did not speak much, was sullen, behaved slightly provoking, and did not keep appointments and agreements. When he was discharged, he appeared much better, then graduated from high school and began to study medicine. When he was 22, he gradually took a turn for the worse. He isolated himself more and more, and was readmitted as an inpatient when he began to consider committing suicide. This time he appeared continually morose, had little perseverance, and monotonously complained of physical pain in the back of the head that was medically not confirmed. He was constantly unsatisfied and ill-tempered. He unwillingly took part in therapeutic programs, but in contrast would spend hours making wooden animals, which he had learned in occupational therapy. His stay lasted almost two years and he was discharged without much improvement to a therapeutic group home. He did not continue his studies at the university. *At the time of the follow-up study* (when he was 25), he delivered newspapers twice a week. From this he earned enough to pay his rent, which he said was all he needed, as he did not do much. He talked monotonously and appeared sullen and disgruntled; however, he was not unfriendly. He indicated that he constantly felt physical pain at the back of his head. He tried to ease the pain by "rubbing over it".

He spent his time making simple wooden animals. This was all that interested him; his family including his twin brother did not seem to interest him.

M 30–1 was 25 years old at the time of the interview, married with two children (twins), and just completed his first set of exams for his degree in medicine. He showed normal behavior and had never been in psychiatric care.

Family history. The 54 year old mother was personally examined. She was talkative and appeared very concerned about her ill son. No psychological conspicuousness was evident and no psychiatric care had ever been given. The 54 year old father was described as cool and unapproachable; however, he had never been seriously ill. A sister of the father had suffered from epilepsy. Another sister of the father had committed suicide for unknown reasons and a third sister of the father had twice attempted suicide after her divorce.

Diagnoses. *According to DSM-III-R residual type of schizophrenia, according to ICD 10 schizophrenia simplex.*

Leonhard Classification. *Systematic hebephrenia (eccentric).*

Subjects W 33–1 and W 33–2 (discordant)

W 33–2 became ill at the age of 19. Upon admittance, she was orientated and calm, but made "bizarre and peculiar movements"... "standing still with a rapturous expression, with her hands, arms and body in peculiar positions"... "standing motionless, then grasping about her, stretching and straining, looking for her mother". Acoustic hallucinations and mistaking people were evident. Later on, she lay in bed, moving her arms constantly and "swayed her chest and head with a certain gracefulness back and forth". At the same time, she incessantly spoke words that she arbitrarily picked up from her surroundings. Subsequently, she was "very wound up, with extreme motor activity", "throwing slippers and pillows about, pouring coffee on the floor, swearing vulgarly at others but appearing more saucy than malicious". After four months, full remission occurred. Three years later, she became ill again. Abnormal psychomotor activities, disturbances in thought and affect and occasional mistaking of people occurred again: "She tosses around in bed, making sounds that seem to express something pleasant, stares into the corner of the room, occasionally screams loudly and cheers, and from time to time talks to those present as if they were acquaintances". She was healthy when discharged three months later. At the ages of 24, 37, 47, 51, 56 and 61, further hospitalizations were necessary. The following are excerpts of typical psychopathological findings: "puffing and panting in a mannered way, makes herself totally stiff, tries to let herself fall, clownishly and foolishly jumps up when touched" ... "peculiarly restless, appears perplexed and anxious" ... "she has been having pains in her abdomen day and night, feels drawn to the ground, is afraid that her husband is a mass murderer and smells decaying corpses" ... "appears almost totally apathetic, and then throws herself into the arms of her son, moaning rhythmically" ... "shows restless psychomotor activity, says that she feels electricity under her bed, and sings in a high soprano" ... "the admitting examination is very difficult and is constantly being interrupted by intrusive-uninhibited behavior, her face is expressionless and rigid and her movements are slack" ... "confused pressured speech, psychomotor agitation, slams doors" ... "mistrustfully agitated, is fearful, restless psychomotor behavior, hallucinations are not out of the question". After her last stay up to this point, at the age of 61, the following entry was made: "well-balanced, friendly, shows a broad range of affect, the illness is in remittence". *In the follow-up study* the subject was 64 years old and had been widowed for three years. She lived in a well-kept house, was friendly, open and showed no mental conspicuousness.

W 33–1 was 64 years old at the time of the interview, was married with two children and a housewife. She had never been in psychiatric care. The examinations showed normal psychopathological results.

Family history. No other family member could be examined. The mother was deceased and was said to have been healthy. The deceased father had been hospitalized three times because of a manic-depressive disorder. A sister of the father had hung herself, and had been previously treated for depression (her medical records could not be obtained). The twins had no siblings.

Diagnoses. *According to DSM-III-R and ICD 10, catatonic type of schizophrenia (individual phases lasted longer than a half a year).*

Leonhard Classification. *Cycloid psychosis (motility psychosis with elements of confusion).*

Subjects M 34–1 and M 34–2 (discordant)

M 34–1 was 19 years old when he was admitted to a psychiatric clinic for the first time. He was reserved and stated that he had the impression that others were making fun of him, and that he felt he wasn't really a part of things. He claimed to have been in love with a girl and that she did not understand him. Upon discharge he showed little improvement. He was diagnosed as having a "narcissistically tinged neurotic depression". Two more inpatient stays with similar uncharacteristic symptoms followed, at the ages of 21 and 22. Even though he graduated from high school, he never became gainfully employed. When he was 28, he was admitted against his will, because he stood in front of a woman's house for 16 hours and appeared confused. In the hospital, he stated that he had known that woman for two years. Voices had told him her address. These voices would sometimes insult him horribly and other times encouraged him. The voices had told him to wait for this woman. He also spoke confusedly about the mafia, secret societies and blackmailers. During his stay, he was very aggressive towards the nursing staff and doctors, and continually threatened to hit them. He spoke of a great ghost, through whom he could connect with others. Upon discharge, he was improved, but "an insight into his illness could not be assumed". He was readmitted at the age of 31. He was extremely agitated and irritated and spoke of a scheme that wanted to finish him off. Ideas of reference, intrusion and persecution were evident. During the course of his stay, he was taciturn and unapproachable. His demands to be discharged were finally consented to after 14 days, because he otherwise showed normal behavior.

In the follow-up study he was 32 years old and unemployed. He appeared reserved, cautious, but not unfriendly. After a certain level of confidence had been established, he spoke freely about his illness. When confronted with some symptoms recorded in his files (persecutory delusions, voices), however, he became nervous and tense. He said that they were all lies, and that he had been badly treated. Grandiose delusions were now suggested. He stated that the "Xs" had always been something special and stood out among the masses. He did not explain this further, but said that the transition from being a genius to being insane was very slight.

M 34–2 was 32 years old when examined, and in a stable relationship. He had never been in psychiatric care and appeared mentally healthy.

Family history. The 54 year old mother and the 62 year old father were personally examined. The mother was a little frenzied and nervous, and easily lost her train of thought. Psychiatric care had never been given. The father appeared exhausted but otherwise showed appropriate behavior. According to M 34–2, he was easily irritated and excitable. A younger brother of the father had been treated as an inpatient because of alcohol problems. An older brother of the father had committed suicide. The twins had no siblings.

Diagnoses. *According to DSM-III-R and ICD 10, paranoid type of schizophrenia.*

Leonhard Classification. *Affective paraphrenia.*

Subjects M 39–1 and M 39–2 (discordant)

M 39–1 showed signs of mental abnormalities for the first time when he was 25. Concentration and sleep disorders and nightmares occurred. He began to hear voices, and believed that medication was being put in his food. He believed that everyone knew all about him. The paranoid components became less prominent, whereas the acoustic hallucinations continued. At 27 years he was admitted as an inpatient and was examined *within the framework of the study.* He showed a normal range of affect, was somewhat dysphoric but gave information readily. He reported that he had been hearing unpleasant voices for more than one year. The voices would talk about him condescendingly and said embarrassing and unpleasant things. He did not like to talk about it. He was able to converse with the voices; they were very annoying and unpleasant. During the cognitive test he did not stick to the point, but often spoke vaguely around the subject. Shortly before he was discharged, he indicated that the voices had receded a bit into the background. However, a tendency towards dissimulation was unmistakable. He was readmitted one year later with the same symptoms. This time also, the voices had not completely disappeared before he was discharged. Despite having a university degree in pharmacy, he was unable to become employed. He was beginning to think about becoming a nurse.

M 39–2 was 27 years old at the time of the examination, single, employed and had just gotten his high school diploma by going to night school. Psychiatric care had never been necessary and the psychopathological results were normal.

Family history. The 56 year old mother was personally examined. She appeared emotional and slightly anxious, but otherwise showed normal behavior. The father had died of cancer at the age of 46. He had always been mentally healthy. A sister of the mother's mother had died in a psychiatric hospital. Further details were not possible to obtain. A six years younger brother and a three years older brother were healthy.

Diagnoses. *According to DSM-III-R and ICD 10, paranoid type of schizophrenia.*

Leonhard Classification. *Systematic paraphrenia (phonemic).*

Subjects W 44–1 and W 44–2 (discordant)

W 44–1 became ill for the first time at the age of 27 following the onset of her twin sister's serious illness (cerebral apoplexy). She felt responsible for her sister's household and blamed her twin's husband for the illness. She suddenly felt threatened and persecuted, and heard frightening voices. She was examined several times as an outpatient by a psychiatrist, when she spoke of unrequited love for a man. After weeks of loss of appetite and weight, she was finally admitted as an inpatient. She had become very much afraid that her family wanted to abandon her. Upon admission, she was occasionally stuporous and very uncommunicative. She appeared confused, explaining that she was mixed up, helpless and unable to act. She reported that she heard the voice of her boyfriend, and was not able to sort out her thoughts anymore. Her mood ranged from being perplexed and depressed to being anxious-dysphoric and agitated. She was stable upon discharge. She was readmitted six months later, and complained helplessly of loss of weight, physical pains (despite medical reassurance) and feelings of inferiority. Upon discharge, she was improved and able to work. In the following years, three more inpatient stays were necessary. Depressive and irritated-maniacal affect, increased or reduced psychomotor activity, feelings of extreme anxiety with persecutory elements were present. She was well remitted upon discharge. *At the time of the follow-up study,* she was friendly, responsive and showed no conspicuous psychopathological behavior.

W 44–2 was born with a heart defect. Overall, she developed normally physically and psychologically. At 27 years, after an embolism, she suffered a left-sided cerebral apoplexy and was paralyzed on one side. *During the follow up study,* the paralysis had diminished to a large extent. She was mentally healthy and had never been in psychiatric care.

Family history. The mother could be personally examined. She was healthy and had never been in psychiatric care. The father had died of encephalitis at the age of 35. The four years younger brother had been in outpatient psychiatric care because of depression. Two older brothers (nine and 13 years older) and a ten years older sister were mentally healthy. A sister of the mother had been in inpatient care because of an endogenous depression.

Diagnoses. *According to DSM-III-R/ICD 10, schizoaffective psychosis.*

Leonhard Classification. *Cycloid psychosis (anxiety psychosis with elements of confusion).*

Subjects W 45–1 and W 45–2 (discordant)

W 45–1 was 42 years old when she first became ill. She developed an erotomanic delusion, and believed that she was connected to a man through listening devices, and waited in erotic suspense to be met by and then married to him. She was then admitted to the hospital. She was stable upon discharge. Four subsequent inpatient stays followed with similar symptoms. In addition to the erotomanic delusion ideas of anxiety also occurred; for example, she feared that she would be gassed. Her last inpatient stay had occurred at 59 years. Upon discharge, she was affable and friendly and was in a stable psychological condition. *In the follow-up study* she was living in her own clean apartment. She satisfactorily showed insight of former delusional contents, was friendly and appeared mentally normal. She spontaneously accepted her former symptoms as an indication of her disorder.

W 45–2 had never been in psychiatric care and was found to be mentally normal during *the examination.*

Family history. The parents were both deceased. The mother had been healthy until her 65th year. She then suffered a "stroke" and thereafter was repeatedly in inpatient care because of depressions. She died at 72 years. The father died of cirrhosis of the liver at the age of 59. He was said to have always been full of life. Two younger brothers (one and three year(s) younger) and two younger sisters (nine and 15 years younger) were all married and mentally healthy. Two older brothers (two and four years older) were also said to be mentally healthy. On the maternal side of the family, some relatives were said to be rather peculiar. A sister of the mother reportedly exhibited arrogant, spiteful and incomprehensible behavior. A sister of the grandmother was also said to be strange. A daughter of another sister of the mother was reportedly angry, spiteful, constantly giving orders, and under nervous strain.

Diagnoses. *According to DSM-III-R/ICD 10, delusional disorder (erotomanic delusion).*

Leonhard Classification. *Cycloid psychosis (anxiety-happiness psychosis).*

Subjects M 46–1 and M 46–2 (discordant)

M 46–1 became ill at the age of 24 years. He complained of physical ailments such as headaches, dizziness, gastritis and sleep disorders. Phobias (fear of heights, agoraphobia), derealization- and depersonalization phenomena existed. These symptoms abated to a great extent without psychiatric care. At the age of 28, delusions of reference and meaning suddenly occurred. He believed that radio programs were transmitted for the sole reason of confusing him. He considered himself to be Jesus and donated a lot of his money to the church, out of proportion to his means. He heard commanding and commenting voices, and claimed an absolute faith in God along with a fear of hell. He was depressed and complained of feelings of insufficiency and failure. The inpatient stay lasted over a year. After this he was discharged in a sufficiently stabilized state. *In the follow-up study* he was 29 years old and enrolled as a student. He supported himself by doing odd jobs. He appeared soft hearted, shy and a little anxious. Residual mental symptoms were not evident. He showed a broad range of affect, was friendly, and very cooperative.

M 46–2 was employed and married. He had never been in psychiatric care and proved to be mentally healthy *during the examination.*

Family history. The 63 year old father, the 62 year old mother, and the two years older sister could all be personally examined. They were all healthy and had never been in psychiatric care.

Diagnosis. *DSM-III-R/ICD 10, schizoaffective psychosis.*

Leonhard Classification. *Cycloid psychosis (anxiety-happiness psychosis).*

Subjects M 47–1 and M 47–2 (discordant)

M 47–2 became ill when he was 30 years old. He believed that his co-workers were saying bad things about him. He also believed that someone was reporting him to the police and punishing him. He asked the mayor several times whether something of that nature had occurred and said he was "an absolute nervous wreck". After he was treated as an outpatient, he felt better. At 34 years, he was admitted as an inpatient. He reported that people were trying to ruin him by saying bad things about him. Whenever he saw people standing about in a group, he felt that they were talking about him. He was very much afraid that by these means he would lose his job. His worry and anxiety became so great that he came to the clinic. During his inpatient stay, several anxious-paranoid states occurred. Upon discharge he was well improved and resumed his work. *In the follow-up study* he was employed and lived with his parents. He appeared a little laggard and complained of fatigue. Since his inpatient stay, he has taken clozapine regularly (150mg daily). He showed a broad range of affect and demonstrated insight into his illness. Psychopathological residual symptoms were not evident.

M 47–1 was employed, married with two children and had never been in psychiatric care. He was stable and mentally normal *during the examination.*

Family history. The 75 year old father and the 73 year old mother could be personally examined. Both parents were healthy and had never been in psychiatric care. A seven years older sister was also mentally healthy.

Diagnosis. *According to DSM-III-R, schizophreniform psychosis, according to ICD 10, acute transient psychotic disorder.*

Leonhard Classification. *Cycloid psychosis (anxiety psychosis).*

Summary of the family history

Table 24 shows how many of the families of the 22 *monozygotic pairs* (8 discordant pairs, 12 concordant pairs C1 + C2, 2 concordant pairs C3) also have other first- and second-degree relatives which have displayed psychiatric symptoms. In the families of the 12 concordant pairs, other psychoses were to be found in first- and second degree relatives in 41.7% (concordance group C1 + C2)

Table 24. Number of families of the 22 monozygotic pairs with other first- and second-degree relatives which have displayed psychiatric symptoms compared with concordant and discordant pairs (percentages in parentheses)

Type of loading	Concordant C1+C2 n = 12	Concordant C3 n = 2	Discordant n = 8
Family case history without suicides and psychoses	5 (41.7%) 6 (42.9%)	1	7 (87%)
Suicides without psychiatric diagnoses	2 (16.6%)	0	0
Psychoses and suicides with psychoses	5 (41.7%) 6 (42.9%)	1	1 (13%)

and 42.9% (concordance group C1 + C2 + C3). If suicides without a previous psychiatric diagnosis are added, the percentage of the families with a mental illness increases to almost 60%. A familial pattern of suicides and psychoses therefore occurred in more than half of the concordant monozygotic pairs. In the families of the 8 discordant pairs another psychosis occurred in only one family (13%).

The familial loading of the 22 monozygotic pairs after distributing it among the different Leonhard diagnoses is as follows: *Of 11 monozygotic pairs with cycloid psychotic index-twins*, eight pairs are discordant and three pairs are concordant (concordance group C1). Among the diagnoses of the index-twins of the eight discordant pairs there were five subjects with anxiety psychoses, one subject with an anxiety-happiness psychosis and two subjects with confusion psychoses. In one of these eight families the mother suffered from a phasic endogenous depression. The subjects of the three concordant pairs all suffered from motility psychoses. One of these three families had another member (sibling) with an endogenous depression.

The 11 monozygotic pairs with unsystematic schizophrenic index-twins are distributed among six pairs with periodic catatonia, four pairs with affective paraphrenia and one pair with cataphasic index-twins. Of these 11 pairs, nine pairs belong to concordance group C1 and the remaining two pairs to concordance group C3. *There was no clearly discordant monozygotic pair in the unsystematic schizophrenia group.* In four pairs (36.4%) the family case histories were to a large extent normal. In the remaining seven families (63.6%), there were sudden suicides and psychoses in first- and second-degree relatives. All of these psychoses belong diagnostically to the schizophrenic spectrum.

Table 25 shows how many of the *25 dizygotic pairs* (18 discordant pairs, 4 concordant pairs C1 + C2, 3 concordant pairs C3) had evidence of other first- and second degree relatives which have displayed psychiatric symptoms. Half of the family case histories of the 18 discordant pairs are without suicides and psychoses, and unexpected suicides (two cases, 11.1%) and psychoses (seven cases, 38.9%) occurred in the other half. Thus, in the case of the dizygotic pairs, 50% of the discordant pairs had a positive family case history with unexpected suicides and psychoses. The families of the four concordant pairs of concordance group C1 + C2 showed no illness in one case and in three cases showed other psychoses. In three pairs of concordance group C3 an unexpected suicide occurred in two families in second-degree relatives and in one family there were other psychoses in first-degree relatives.

The familial loading of the 25 dizygotic pairs after being distributed among various Leonhard diagnoses is as follows: *Of 11*

Table 25. Number of families of the 25 dizygotic pairs with other first- and second-degree relatives which have displayed psychiatric symptoms in comparison with concordant and discordant pairs (percentages in parentheses)

Type of loading	Concordant C1 + C2 n = 4	Concordant C3 n = 3	Discordant n = 18
Family case histories without suicides and psychoses	1	0	9 (50%)
	1 (14.3%)		
Suicides without psychiatric diagnoses	0	2	2 (11.1%)
	2 (28.6%)		
Psychoses and suicides with psychoses	3	1	7 (38.9%)
	4 (57.1%)		

dizygotic pairs with cycloid psychotic index twins, nine pairs were discordant and two pairs were concordant. The diagnoses of the nine subjects of the discordant pairs were distributed among one motility psychosis, one confusion psychosis, four anxiety-happiness psychoses and three anxiety psychoses. In four of these families the family case histories were normal. In three families with index twins suffering from anxiety psychoses, other immediate relatives suffered from phasic depressions. In the family with the index twin suffering from confusion psychosis, the mother suffered from postpartum depression after the birth of her first child and the father of the mother committed suicide (no psychiatric diagnosis known). In the family with the index case of a motility psychosis, the father suffered from a manic-depressive disorder and the sister of the father committed suicide in the course of an endogenous depression.

In one concordant pair both subjects suffered from anxiety psychoses and in the other concordant pair both subjects suffered from confusion psychoses. The rest of the family case history of the former pair was normal. In the family of the other pair, a sister and a brother both suffered from an endogenous depression.

Of the three dizygotic pairs with periodic catatonic index-twins, one pair was discordant, one pair was concordant C1 and one pair was concordant C3. In all three families there were other psychoses in first- and second-degree relatives. In one family the father suffered from a paranoid-hallucinatory psychosis, in another family the mother and two brothers also suffered from periodic catatonia, and in the third family the father committed suicide after suffering for many years from a chronic paranoid depression.

The five pairs with affective paraphrenic index twins were distributed among two discordant pairs, two concordant pairs of concordance group C3 and one concordant pair of concordance group C1. In the family of one discordant pair there was no knowledge of suicides or psychoses. In each of three families there had been a suicide (no psychiatric diagnoses known) of a first- or second-degree relative and in one family the mother suffered from phasic depressions.

All the six dizygotic pairs with systematic schizophrenic index twins were discordant. In four families there was no evidence of illness in the rest of the family history. In one family a distant relative died in a psychiatric hospital (no diagnosis known). In another family there were cases of suicide in second-degree relatives.

Summary of monozygotic and dizygotic pairs: In 18 of the 47 families (38.3%) there were other psychoses in first- and second-

Table 26. Family case histories of the 47 twin pairs (monozygotic and dizygotic pairs after the diagnostic distribution of the index subjects (*n* number of first-degree relatives)

Diagnoses of the index twins	First-degree relatives with			
	psychoses of schizophrenic spectrum	affective psychoses	suicides without psychiatric diagnoses	other psychiatric symptoms
Total sample (193 relatives)	4.2%	5.2%	1.6%	13.5%
DSM-III-R/ICD10 schizophrenia (73 relatives)	8.2%	2.7%	4.1%	15.1%
Other spectrum psychoses (120 relatives)	1.7%	6.7%	–	12.5%
Unsystematic schizophrenia (65 relatives)	12.3%	4.6%	4.6%	29.2%
Systematic schizophrenia (22 relatives)	–	–	–	13.6%
Cycloid psychoses (106 relatives)	–	6.6%	–	3.8%

degree relatives. In six families, suicides occurred without a psychiatric diagnosis (12.8%). Table 26 gives a summary of the familial loading of all pairs for first-degree relatives after the diagnostic distribution of the index cases. Personality disorders, alcohol abuse and other long-term mental disorders are summarized in the column headed "other psychiatric symptoms" with the exception of demential disorders and mental retardation.

Birth history in the intra-pair comparison

In each of the following cases the ill partner is compared with the healthy partner in the discordant pairs, and the more severely ill partner is compared with the less severely ill partner in the concordant pairs. The decision as to who was the more severely ill subject in a concordant pair was made considering the age at the onset of the disorder, the number and duration of hospitalizations, social competence in the follow-up study and the overall clinical impression made during the follow-up study. The data originates from the case histories of the index subjects and from the personal interviews of all subjects, 38 mothers, 21 fathers and four siblings who were considerably older than the subjects. Seven pairs had no remaining living relatives. In three of these seven pairs (W9, W22, W33) no details of the birth weight were available in the medical records, and in the case of one pair (W33) there were no details of possible perinatal complications. Overall, data was obtained on the birth weight of 43 pairs (92%) and on information of perinatal complications in 45 pairs (95.7%). Tables A8 and A9 in the Appendix list the data obtained retrospectively on the birth history of the subjects.

No differences in birth weight were found between healthy/ill or less ill/more ill subjects either in the overall test groups of monozygotic and dizygotic pairs or in the individual diagnostic subgroups. No significant differences were to be found in the order of delivery either; i.e. statistically, the ill or more severely ill subject was the first or second born child with an equal degree of frequency. *In the case of the monozygotic subjects* the ill or more severely ill subjects had significantly more frequent and serious perinatal complications than their partners in the overall group (schizophrenic spectrum), in the group "other diagnoses of the schizophrenic spectrum apart from schizophrenia according to DSM-III-R/ICD 10", and in the group of cycloid psychoses. *In the case of the dizygotic subjects* this finding was evident in the group of the DSM-III-R/ICD 10 schizophrenia (Tables 27 and 28). There was no differ-

ence in the average number and severity of perinatal complications compared with the individual diagnostic categories. The diffe-

Table 27. Number of perinatal complications: the ill or more severely ill subject had perinatal complications more frequently than his healthy or less ill partner (Wilcoxon matched-
pairs signed rank test)

Diagnoses	Monozygotic pairs	Dizygotic pairs
Schizophrenic spectrum	p < .01[1]	ns
Schizophrenia (DSM-III-R/ICD10)	ns	p < .01[1]
Other diagnoses of the schizophrenic spectrum (DSM-III-R/ICD10)	p < .01[1]	ns
Unsystematic schizophrenia	ns	ns
Systematic schizophrenia	–	n too small
Cycloid psychoses	p < .01[1]	ns

[1] Significant also according to the Bonferroni-correction.

Table 28. Severity of perinatal complications: the ill or more severely ill subject had perinatal complications more frequently than his healthy or less ill partner (Wilcoxon matched-
pairs signed rank test)

Diagnoses	Monozygotic pairs	Dizygotic pairs
Schizophrenic spectrum	p < .01[1]	ns
Schizophrenia (DSM-III-R/ICD10)	ns	p < .01[1]
Other diagnoses of the schizophrenic spectrum (DSM-III-R/ICD10)	p < .01[1]	ns
Unsystematic schizophrenia	ns	ns
Systematic schizophrenia	–	n too small
Cycloid psychoses	p < .01[1]	ns

[1] Significant also according to Bonferroni-correction.

rences occurred only in the intrapair comparison within a diagnostic category.

Role assignment in the twin constellation

In the case of three pairs (M8, W9, W43) no usable information could be obtained on the role assignment in the pair. Clear information was given in the case of 40 pairs (85%) as to which of the twins dominated or was subordinate during childhood and before the onset of the illness in the pair constellation. Both partners were described as equal in the case of six pairs.

When considering the sample as a whole, the ill or more severely ill subjects both in the monozygotic as well as dizygotic pairs were subordinate to their partner significantly more often (monozygotic pairs: $x^2 = 14,0$, df = 2, p < .01; dizygotic pairs: $x^2 = 16,3$, df = 2, p < .001). After the diagnostic differentiation and Bonferroni-correction, however, the differences were no longer statistically significant.

Handedness of test persons

The frequency of individuals in the general population who are not exclusively right-handed is said to be between 8% and 10% (Oldfield 1971, Luchins et al. 1980). The occurrence of non-right-handed people (NRH) is said to be equally frequent in monozygotic and dizygotic twins as well as in single births (McManus 1980). The handedness of all subjects is shown in Tables A10 and A 11 of the Appendix. 14 of the 94 subjects (14.9%) were non-right-handers. There were no major differences in the frequency of NRH in monozygotic (five of 44 subjects = 11.4%) and dizygotic twins (nine of 50 subjects = 18%), so it does not seem appropriate to analyze handedness separately according to zygosity. In the sample as a whole and in the various diagnostic subgroups, there was no considerable difference in the percentages of NRH between healthy and ill subjects. The proportion of pairs with a NRH was higher for the discordant pairs than for the concordant pairs in the sample as a whole and in each of the diagnostic subgroups. However, the difference was not statistically significant.

Discussion

Despite the enormous progress made in investigative methods in basic sciences, social- and psychodynamics and in epidemiological and social-psychiatric research, increasing stagnation has occurred in acquiring of knowledge of the psychoses of the "schizophrenic spectrum" during the last several years. It has still not yet been settled whether the spectrum of schizophrenic and schizophrenia-like psychoses is a disease continuum with continuous boundaries or whether it is composed of different diseases with very different causes. The great hope placed in modern, atheoretical and operational diagnostic systems has not yet been realized. Further, there are very few findings which have not been contradicted. It therefore seems justified to question whether the mere compilation of a number of specific symptoms based on expert consensus, in order to form diagnostic categories and/or "disorders" without a nosological background, is actually sufficient to advance schizophrenia research. This expert consensus is a compromise of various individual viewpoints and should serve first of all to improve mutual understanding. In attaining a high degree of diagnostic correspondence between different investigators (reliability), however, it is often necessary to pay the price of over-simplification. It is also incorrect to automatically assume that the consensus criteria are also valid. *Reliability does not necessarily mean validity* (Gottesman and Shields 1982). A systematic twin study which compares diagnostic systems based on "expert consensus" (DSM-III-R, ICD 10) with Leonhard's nosology based on clinical-empirical research appears to be appropriate for checking the validity of diagnostic subgroups within the schizophrenic spectrum.

Discoveries in the biology of twin origination and the confirmed effects of intrauterine environmental differences in about two-thirds of the monozygotic twins with regard to their prenatal development have made it necessary to extend investigation to a comparison of the concordance rates between monozygotic and dizygotic twin pairs. The transfusion syndrome in 15% to 30% of monochorionic monozygotic twins alone could possibly result in either increased or decreased concordance rates in primarily hereditary as well as primarily non-hereditary diseases. For this reason we took on Campion and Tucker's appeal (1973) to give pre- and perinatal complications more consideration in twin studies, and have given special attention to pregnancy and birth histories, birth weight and birth order. The retrospective survey of these data has been assessed in international studies as being sufficiently valid (Wenar and Coulter 1962, National Center for Health Statistics 1985, Little 1986, Gayle et al. 1988, O'Callaghan et al. 1990a). Based on a very precise survey of family histories, of demographic data taken at the time of the onset of the illness and the follow-up examination, the determination of the pre-psychotic role assignment in the pair constellation (dominant/subordinate axis) and handedness, we expected to obtain further information in order to answer our questions. The discussion of the findings obtained *from 47 twin pairs (22 monozygotic and 25 dizygotic pairs)* is carried out on the basis of results from the major twin studies on schizophrenia conducted thus far.

Role assignment (dominant/subordinate) and handedness

Our findings on the role assignment within the pair constellation confirm the results of earlier twin studies. Among all psychoses of the schizophrenic spectrum, the ill (or more severely ill) twin was often subordinate to his partner at the pre-psychotic stage. Like Kringlen (1990), we were able to record this finding among monozygotic as well as dizygotic pairs. After dividing the schizophrenic spectrum into diagnostic subgroups, however, statistically significant differences could no longer be detected, making new conclusions impossible.

The investigations of handedness showed no statistically significant differences in the frequency of "non-right-handed" subjects in certain diagnostic subgroups, neither when comparing monozygotic and dizygotic pairs nor when comparing concordant and discordant pairs. Like Lewis et al. (1989) and Torrey et al. (1993a), we were also unable to confirm the hypothesis of Boklage (1977) that left-handedness should not be associated with "nuclear schizophrenia", but rather with a more favorable form of the disorder.

Comparability of diagnoses

It is often impossible to compare results if different diagnostic concepts are used in the different studies. The study by Gottesman and

Shields (1972) is a striking example. Among six different diagnosticians the frequency of the diagnosis of schizophrenia fluctuated
between 44 and 77 in a group of 114 test persons (Table 4). We
therefore analyzed the diagnoses of the 64 psychotic subjects in
terms of determining whether there were parallels between the
DSM-III-R, ICD 10 and the Leonhard classification systems.

All systematic schizophrenias according to Leonhard within our
study also fulfilled the criteria for schizophrenia according to
DSM-III-R/ICD 10. In contrast, unsystematic schizophrenia and
cycloid psychoses according to Leonhard were distributed in the
DSM-III-R/ICD 10 over a number of diagnoses. The DSM-III-R/
ICD 10 diagnosis of a schizoaffective disorder proved to be particularly inconsistent and was inconclusive for making a specific
Leonhard diagnosis. Both cycloid as well as unsystematic schizophrenic psychoses were found in approximately equal proportions.
Fairly reliable conclusions with regard to a specific Leonhard diagnosis, however, appeared to be made in the categories "schizophreniform disorder" (DSM-III-R) and "acute transient psychotic
disorders" (ICD 10). In our test group, only subjects with cycloid
psychoses were found in both categories.

Demographic data in polydiagnostic comparison

In the assessment of demographic data it became apparent that both
the monozygotic as well as the dizygotic female pairs were significantly older than the male pairs at the time of the investigation.
Further analysis of the data showed that these findings were attributable to the fact that the female test persons became ill on average
somewhat later than the male test persons. The later onset among
schizophrenic women compared to schizophrenic men, observed
by many researchers (Loranger 1984, Flor-Henry 1985, Häfner et
al. 1991, Franzek and Beckmann 1992b), was also found in our
twin test group with schizophrenic spectrum psychoses. After
breakdown into different diagnostic categories only slight differences were apparent in age at onset of the disorder between the
sexes, and only in the large category "other diagnoses of the schizophrenic spectrum except schizophrenia" (DSM-III-R/ICD 10)
and in the Leonhard category of cycloid psychoses. In polydiagnostic comparison of the three classification systems, cycloid psychoses were found in this "other spectrum diagnoses" with only
one exception, and female pairs were found with cycloid psychoses more often than male pairs. In our twin test group the higher
age of the onset in women over men is attributable to the female

test subjects with cycloid psychoses. As it is known that cycloid psychoses often become manifest in the postpartum period (Pfuhlmann et al. 1998), it would be possible to speculate that the influence of estrogen on the age of onset in women assumed by some authors (Seeman 1982, Häfner et al. 1991) is particularly relevant to cycloid psychoses. Of the schizophrenic patients in our test group, there were no differences in the age of onset between the men and women. In an earlier investigation as well we had found no age differences in the subgroups of schizophrenia (according to Leonhard) with regard to the onset of illness among men and women. We determined, however, that women were diagnosed with paraphrenia more often than men and men with catatonia more often than women (Franzek and Beckmann 1992a). These findings were also partially confirmed in our twin test group. Considerably more men than women were diagnosed with catatonia, whereas the proportion of women to men with paraphrenia was about equal.

The results of this study on the social situations of all the twin test persons confirm the findings of Kringlen; namely, that pre-psychotic social differences can be found in both discordant monozygotic as well as discordant dizygotic twins (Kringlen 1967, 1990). The division of ill subjects into diagnostic subgroups revealed considerable differences, however. Compared with test persons with systematic schizophrenia, affective paraphrenia and cycloid psychosis, test persons with periodic catatonia had had in many more instances only a limited formal education. Only one-half of the periodically catatonic test persons had an occupation at the time of their first hospitalization (school, apprenticeship or profession), whereas this applied to all test persons with affective paraphrenia and cycloid psychosis. Although one-half of the six twins with systematic schizophrenia had gone on to pursue higher education, none was employed at the time of their first hospitalization. No test person with periodic catatonia and systematic schizophrenia was in a relationship at the time of initial hospitalization, while this applied to more than a quarter of the test subjects with diagnoses of affective paraphrenia and cycloid psychosis. These findings indicate that pre-psychotic irregularities occur more often among the periodically catatonic and systematic schizophrenic. The long-term prognosis was also very poor in every case. None of the test persons with periodic catatonia and systematic schizophrenia was able to live independently in a normal social environment up to the follow-up examination (20 years after the first hospitalization, on average).

In the social outcome there were also considerable differences between test persons with affective paraphrenia and test persons with cycloid psychoses. Of the affective paraphrenic subjects who were still 100% socially integrated at the time of their initial hospitalization, only 38% remained well integrated at the time of the follow-up examination. In contrast, 67% of the subjects with cycloid psychoses were still well integrated at this later time. Our results confirm the especially poor long-term prognosis of systematic schizophrenia described by Leonhard (1995). Within the range of unsystematic schizophrenia, affective paraphrenia appeared to have a more favorable social outcome than periodic catatonia. In the case of cycloid psychoses it was possible to reconfirm a good prognosis over the long-term (Perris 1975, Brockington et al. 1982, Beckmann et al. 1990).

Varying genetic dispositions

Schizophrenia generally continues to be treated as a unified disease or at least a unified disease spectrum, based on Kraepelin's dichotomy of endogenous psychoses in manic-depressive disease and dementia praecox. Despite this, attempts have been repeatedly made to divide schizophrenic psychoses into valid subgroups (Kahlbaum 1863, Wernicke 1900, Kleist 1925, Leonhard 1956, Kay and Roth 1961, Tsuang et al. 1974, Harrow and Quinlan 1977, Strauss et al. 1977, Murray and Murphy 1978, Pope and Lipinsky 1978, Crow 1980, Quitkin et al. 1980, Andreasen and Olsen 1982, Lewis et al. 1987b). In classical twin research there are many findings which question the unity of the schizophrenic spectrum. The often clearly varying concordance rates of the monozygotic twins in the major twin studies also indicate that certain patient subgroups apparently predominate in individual studies, which either raise or lower the concordance rates. For example, studies based on hospital patients often report considerably higher concordance rates than studies which are based on twin registers. Studies which recruited twins from hospitals have been accused repeatedly of the systematic error of investigating primarily severely ill subjects, which are more often concordant than less severely ill test persons, because they show a quantitatively higher genetic loading (Kringlen 1990). In our twin studies as well, the test persons which fulfilled the strict schizophrenia criteria specified by DSM-III-R showed extremely high monozygotic concordance rates and significantly lower dizygotic concordance rates. The test subjects with other DSM-III-R spectrum diagnoses, in contrast to schizophrenia, had a lower concordance rate (showing a statistical trend) among the monozygotic test subjects. The difference in the dizygotic concordance rate was no longer significant in this group. At first glance this appears to confirm the assumption of a quantitatively higher genetic pattern among severely ill individuals compared with mildly ill persons.

Another consideration, however, is the possibility that the so-called "severe illnesses" and "mild disorders" could be etiologically different psychoses. In the final analysis the concept of differentiated diagnostics according to Leonhard is based on this assumption. In using this diagnostic system, extremely high monozygotic concordance rates and significantly lower dizygotic concordance rates among the unsystematic schizophrenic patients were found in our test group. Among the cycloid psychoses, however, the majority of the monozygotic pairs were also discordant and the concordance rate of these monozygotic pairs was only slightly higher than the concordance rate of the dizygotic pairs. Further, the concordance rate of the monozygotic test subjects with unsystematic schizophrenia was (statistically) significantly higher than the concordance rate of the monozygotic test subjects with cycloid psychoses. *According to Galton's law these findings should be interpreted as that unsystematic schizophrenia is primarily genetically determined and that, in contrast, heredity apparently plays only a very subordinate role among cycloid psychoses.*

Affective psychoses were found in nine of 22 families with a cycloid psychotic index-twin (monozygotic and dizygotic). The loading of affective psychoses was 6.6% among a total of 106 first-degree relatives. In contrast there was no case in which a psychosis of the schizophrenic spectrum appeared among first-degree relatives. This shows that cycloid psychoses are much more similar to affective psychoses than to schizophrenic psychoses. The interpretation, however, that cycloid psychoses are simply a variant of manic-depressive illness is certainly not correct, since twin studies on manic-depressive illness report extremely high concordance rates among monozygotic test persons and significantly lower concordance rates among dizygotic ones (Bertelsen et al. 1977), which is significantly different from our results. The findings point instead to the nosological autonomy of cycloid psychoses, although the precise link to purely affective psychoses must still be investigated.

The familial pattern of unsystematically schizophrenic index-twins was qualitatively much different. In 13 of 19 families unexpected suicides or psychoses of the schizophrenic spectrum occurred. Members of only two families suffered from purely affective psychoses. In total 12.3% of the 65 first-degree relatives had a history of psychoses of the schizophrenic spectrum. The loading of other long-term mental disorders (personality disorders, alcohol dependency, etc.) was also very high (29.2%). In the case of periodic catatonia, the findings are compatible with other investigations which assume a dominant genetic factor in the sense of a major gene effect (Beckmann et al. 1996, Leonhard 1995, Stöber et al. 1995). In any case, the family findings and the statistically significant different concordance rates confirm the varying genetic pattern of the group of cycloid psychoses and the group of unsystematic schizophrenia already postulated by Leonhard.

If one includes the cycloid psychoses in the schizophrenic spectrum, our findings also confirm Rosenthal's postulate (1959) that there are two types of schizophrenia: a non-hereditary type with good prognosis (cycloid psychosis, according to Leonhard), and a hereditary type with an unfavorable course (unsystematic schizophrenia, according to Leonhard). Rosenthal had also observed that concordant monozygotic twins show a greater incidence of a family history of schizophrenia and that the course of the disease and its outcome were generally unfavorable. Among the discordant monozygotic pairs he found no positive family histories and the disease usually took a favorable course. In eight of the 14 concordant monozygotic pairs in our study (57%), suicides and psychoses occurred among first- and second-degree relatives. Furthermore, unsystematic schizophrenia was found in 11 of 14 concordant monozygotic pairs (79%), and cycloid psychoses in only three pairs. The family histories of the eight discordant monozygotic pairs showed only one case of a psychosis (endogenous depression), and all index cases of these discordant pairs suffered from a cycloid psychosis. The significance of the absence of systematic schizophrenia among monozygotic twins is discussed below in further detail.

Leonhard had also observed a concordance rate of 56% among his monozygotic twin subjects with cycloid psychoses (10 of 18 pairs were concordant). Nine of his ten concordant pairs belonged to the clinical subgroup of a motility psychosis. In the three concordant monozygotic pairs with cycloid psychoses in our group, the subjects also exhibited the clinical symptoms of motility psychoses. This high concordance rate among motility psychoses stands in contrast to their relatively low familial pattern. Leonhard (1976) therefore assumed a considerable additional somatic influence in the peri- and postnatal phase in addition to genetic disposition. It is likely that pregnancy and birth complications take special priority in their etiology. This will be discussed in greater detail below.

The absence of schizophrenic subgroups among twins

A purely psychological-psychodynamic hypothesis is that the insufficient development of ego boundaries, or "confusion of ego identity" (Bateson et al. 1956, Jackson 1959), increases the likelihood of the development of schizophrenic psychoses. Twins very often show a strong attachment to one another in an attempt to achieve similarity and identification. This attachment is much more pronounced among the monozygotic pairs than the dizygotic pairs. This attachment may possibly hinder the development of an independent and autonomous personality more frequently among monozygotic twins than dizygotic twins, and more frequently among dizygotic twins than ordinary siblings. Without taking genetics into account, it would be possible to derive higher concordance rates among monozygotic twins than among dizygotic twins. Likewise, this would also explain the higher concordance rates of dizygotic twins compared with ordinary siblings diagnosed with the disease (Table 6). According to this hypothesis, however, it would be necessary to establish that

1. schizophrenic psychoses in twins (monozygotic and dizygotic) occur more frequently than in the normal population,

2. schizophrenic psychoses in monozygotic twins occur more frequently than in dizygotic twins.

Neither assumption could be confirmed by systematic twin studies. Schizophrenia in twins occurred no more frequently than in the normal population and the frequency of schizophrenic psychoses among monozygotic and dizygotic twins also did not differ. The "confusion of ego identity" hypothesis therefore becomes very questionable as a significant etiological factor of schizophrenic psychoses. Moreover, some findings even lead to contrary conclusions. Leonhard did not find a single case of systematic schizophrenia among 69 endogenous psychotic monozygotic twins, but found 12 cases of systematic schizophrenia among 47 endogenous psychotic dizygotic twins. Leonhard was not able to provide a biological explanation for this completely unexpected finding, and posed the hypothesis that a person might possibly not become systematically schizophrenic if someone close to him understands him and all his reactions to an extremely high degree. According to Leonhard, it depends on whether the monozygotic twins grow up together. This inevitably creates a close relationship, ensuring sufficient interaction between the individual and environment and allowing a normal development of "psychic systems" (Leonhard 1979). On the other hand, it must naturally follow that a lack of interaction with the environment would prevent the development of "psychic systems", and that such an individual might possibly become systematically schizophrenic at a later stage.

An example of the severe and dramatic mental consequences stemming from a lack of interaction with the environment is "anaclitic depression" in small children (Spitz 1945). Among small children who grew up in orphanages, where they received only irregular attention from the staff, Spitz described withdrawal, rejection of and resistance to the environment, insomnia, poor appetite, apathy, delayed reaction to environmental stimuli, and mental and physical deterioration even to the point of death. It is also now known that insufficient stimulation in the sensitive postnatal developmental phase can lead to improper synapses or even a complete loss of function of an intact but not properly developed neuron system (O'Kusky 1985, Meisami and Firoozi 1985).

Among 22 monozygotic twin pairs with 34 psychotic twins, there was no twin with systematic schizophrenia in our series of investigations, although it did occur in six of 30 psychotic dizygotic twins. This finding is (statistically) highly significant. As Leonhard did not carry out a systematic twin survey, he could have overlooked systematically schizophrenic monozygotic twins. Despite our systematic recruiting of twin test subjects, however, we were not able to disprove Leonhard's findings.

In our test group we were able to find only one twin pair with a cataphasic index-twin, a clinical subgroup of unsystematic schizophrenia. Leonhard provides no information in his publications on twin research with regard to cataphasia. A review of his case histories showed two monozygotic concordant pairs, one discordant pair with uncertain zygosity, and two discordant dizygotic pairs diagnosed with cataphasia. In relation to the total number of 116 psychotic twins, the number of twins with cataphasia (6%) was also relatively small in Leonhard's study. This result is also surprising. The special twin constellation of monozygotic twins cannot be the reason, since this result was also found among the dizygotic twins. Other factors which are associated with twin birth could also play an important role here. These will be discussed below in greater detail.

The influence of pre- and perinatal complications

The significance of pregnancy and birth complications in the origin of schizophrenic psychoses is very controversial (McNeil 1987, Goodman 1988, Stöber et al. 1993a). During the last several years, the issue of whether there is a correlation between preg-

nancy/birth complications and ventricle anomalies and/or the likelihood of genetic disease among schizophrenic patients has been
investigated in particular. A number of authors found a positive
correlation between pregnancy/birth complications and ventricle
dilation among schizophrenic patients (Pearlson et al. 1985, Turner
et al. 1986, Lewis et al. 1987a, Owen et al. 1988, Cannon et al.
1993). Others found no relation (Reveley et al. 1984, Oxenstierna
et al. 1984, Nimgaonkar et al. 1988, Nasrallah et al. 1991), and
still others reported a negative correlation (Kaiya et al. 1989).
Equally contradictory were the results of the studies investigating
the relationship of pregnancy/birth complications and the genetic
likelihood of developing the illness. On the one hand there are
studies which found an increased rate of pregnancy/birth complications among a scattered pattern of schizophrenia as opposed to a
familial pattern of schizophrenia (Wilcox and Nasrallah 1987,
Schwarzkopf et al. 1989, O'Callaghan et al. 1990b, Stöber et al.
1993b); whereas on the other hand, no differences were found
(Pearlson et al. 1985, Nimgaonkar et al. 1988, Reddy et al. 1990).
Conversely, a correlation between a high genetic risk and preg-
nancy/birth complications was reported (Cannon et al. 1993).

There is consensus that certain subgroups of schizophrenic patients show ventricle anomalies (Johnstone et al. 1976, Owens et
al. 1985), which are already present before the onset of the illness,
and therefore not a consequence of it or its treatment (Weinberger
1987, Breslin and Weinberger 1991, Cannon 1991), and that ventricle anomalies are the most frequent consequence of perinatal
trauma (Kulakowski and Larroche 1979, Bergström et al. 1984).
This is confirmed by several neuroradiological twin studies which
found that among monozygotic pairs which were discordant for
schizophrenia, the schizophrenic twins had significantly wider
ventricles than their healthy co-twins (Reveley et al. 1982, 1984,
Suddath et al. 1990).

In this twin study we recorded detailed birth histories and compared the ill and healthy partners of the discordant pairs, and the
more severely ill partner with the less ill partner of the concordant
pairs. This was done separately for the monozygotic and dizygotic
pairs. The results indicate that the contradictory literature on the
influence of pregnancy/birth complications can probably be traced
back to the diagnostic heterogeneity of the examined group of patients. In the case of birth weight and birth sequence, we found no
significant differences in the diagnostic subgroups between the
healthy/ill and more/less ill twin. Our results for birth weight and
birth sequence corresponded with those of other systematic twin

studies which also found no significant differences (Gottesman and Shields 1976, Reveley et al 1984, Lewis et al. 1987a, Onstad et al. 1992).

In some diagnostic subgroups, however, the ill or more severely ill subjects had a significantly higher rate and severity of birth complications than their healthy or less ill partners. Among the monozygotic pairs this was the case in the whole test group and in the subgroups of "other diagnoses of the schizophrenic spectrum" (DSM-III-R/ICD 10) and "cycloid psychosis" (according to Leonhard). Since most monozygotic test subjects with cycloid psychoses are found in the category "other diagnoses of the schizophrenic spectrum", the significant results in this subgroup (and also in the overall group of the schizophrenic spectrum) are attributable to the mainly discordant cycloid psychoses. Franzek et al. (1996) also reported that patients with cycloid psychoses showed neuroradiological indications of early childhood brain damage significantly more often than patients with other endogenous psychoses. *Pregnancy/birth complications therefore appear to play a significant etiological role in cycloid psychoses.* Another important finding is that the average total number and severity of complications among the monozygotic test subjects with cycloid psychoses was not greater than among the test subjects with other diagnoses, and that there were significant differences only in the intrapair comparison with the co-twin.

Among the dizygotic pairs, in the diagnostic subgroup "schizophrenia" according to DSM-III-R/ICD 10, the ill (or more severely ill) test persons had a statistically significant higher rate and severity of birth complications than their co-twins. The exact analysis of the data shows that this result was found mainly in the exclusively dizygotic systematic schizophrenic patients: the systematic schizophrenic patients were all dizygotic and all fulfilled the DSM-III-R/ICD 10 criteria for schizophrenia. Altogether, the six systematic schizophrenic patients had an average of three times as many and three times more severe complications than their healthy partners. In contrast, no intrapair differences were found among the dizygotic unsystematic schizophrenic patients. The DSM-III-R/ICD 10 group "schizophrenia" of the dizygotic twins consists only of unsystematic and systematic schizophrenia. This reinforces the findings of Stöber et al. (1993ab) that *pregnancy/birth complications play an important etiological role in systematic schizophrenic patients.*

The role of prenatal developmental disorders

Twin studies (Morison 1949, Price 1950, Sydow and Rinne 1958, Rausen et al. 1965, Bulmer 1970, Campion and Tucker 1973) as well as research on cerebral palsy (Kuban and Leviton 1994) emphasize that birth complications are only one factor in a continuum of possible impairments which can occur during the entire course of pregnancy, birth and the early postnatal period. Birth complications are frequently only the consequence and therefore the epiphenomena of pregnancy complications (Nelson and Ellenberg 1986).

In a postmortem study, Jakob and Beckmann (1984, 1986) described for the first time cytoarchitectonic changes in the entorhinal region of the parahippocampalis gyrus in the brains of schizophrenic patients. They interpreted this discovery as a disorder of the neuron migration in the maturing brain of the fetus. Several research groups have reproduced this discovery since (Falkai et al. 1988, Arnold et al. 1991) and have also found this in other regions of the brain (Akbarian et al. 1993). The normal migration of neurons from the ventricular zone to their predetermined positions in the central nervous system takes place almost exclusively during the second trimester of the pregnancy. Twin studies have also resulted in indirect indications that a developmental disorder of the brain in the second trimester could be related to the development of schizophrenic psychoses. Among monozygotic twins which were discordant for schizophrenia, significantly more minute malformations of the hands were found in the ill twins. These were traced back to a disorder in the migration of ectodermal cells in the upper limbs (Bracha et al. 1991, 1992). Nerve and skin cells originate from the same germ layer (ectoderm), and the migration of cells which form the skin of the hands also takes place during the second trimester of pregnancy. This could mean that in the case of schizophrenic psychoses the brain development disorder during the second trimester is primary, and that other additional perinatal complications often follow secondarily.

Epidemiological investigations indicate that brain development disorders could also be affected by harmful environmental factors. It was found that the birth rates of schizophrenic patients always increased when the mothers were exposed to epidemics of infectious diseases (usually viral epidemics) during the second trimester (Mednick et al. 1988, Torrey et al. 1988, O'Callaghan et al. 1991a, Barr et al. 1992, Franzek and Beckmann 1992c, Sham et al. 1992, Adams et al. 1993). The repeatedly verified surplus of birth rates of schizophrenic patients in the winter and early spring months also supports this line of thought (Bradbury and Miller 1985). Viral infections occur more frequently when it is colder and, if the mother becomes ill, could also damage the fetus directly or indirectly (Torrey et al. 1988). Some studies have investigated the seasonality of the birth rate of schizophrenia patients in relation to the familial pattern of psychoses, and have determined that a high birth rate in winter/spring months was found only among schizophrenic patients without a familial pattern (Kinney and Jakobsen 1978, Zipursky and Schulz 1987, D'Amato et al. 1991, O'Callaghan et al. 1991b). Franzek and Beckmann (1992c, 1996) also applied the Leonhard classification in a study on the seasonality of births of schizophrenic patients. They found that in a representative test group the high birth rate was achieved exclusively by patients with cycloid psychoses and systematic schizophrenia (with a low genetic pattern), while patients with unsystematic schizophrenia (high genetic pattern) showed a slight decrease in the birth rate in the relevant months. In the case of cataphasia (a clinical subgroup of unsystematic schizophrenia) the decrease in birth rate was statistically significant. This could possibly mean that damaging environmental factors are important etiologically only among psychoses with a low or no genetic pattern, while exogenous stressors play only a subordinate etiological role among psychoses with a high genetic pattern. It is also plausible that with a given genetic disposition to an illness, the addition of exogenous stressors can increase the manifestation of the illness (Cannon et al. 1993). Stöber et al. (1993) reported that among the genetically high risk unsystematic schizophrenics, birth complications led to a premature onset of the illness. In a study of birth seasonality it would therefore be expected that especially unsystematic schizophrenia would have a pronounced higher birth rate in winter/spring months when genetic disposition and predisposing environmental stressors occur together. How then can the opposite findings, a lower birth rate, be explained?

Animal experiments and neuropathological findings have shown that the migration of neurons can be disturbed both by exogenous stressors as well as by genetic defects (Rakic 1988). Beckmann and Jakob (1991) reported that the neuron migration disorder is probably a general vulnerability factor of "functional psychoses" and can also be found among psychoses with a predominant genetic loading. Beckmann and Franzek (1992) thus derived the hypothesis that exogenous stressors at a critical point in the development of a brain already damaged by a genetic defect, can as a lethality factor lead to miscarriage, stillbirth or premature death in childhood. Individuals with the genetic disposition to cataphasia could possibly be at especially high risk in such a case. This may explain the rare occurrence of these psychoses in our twin group. A twin pregnancy and birth is already viewed as a complication (see Table 10). There are a number of investigations that reported an increased rate of miscarriages and stillbirths, as well as an increased rate of infant and child mortality among the children of schizophrenic parents (Kallmann 1938, Sobel 1961, Videbech et al. 1974, Rieder et al. 1975, McSweeny et al. 1978, Modrewsky 1980). Studies by Torrey et al. (1993b, 1996) also support this hypothesis. The authors reported a significantly high birth rate of schizophrenics born in the winter months and a significantly high rate of stillbirths in this time period, as well as a higher rate of stillbirths in every month from January to June and a significant correlation of stillbirths to the occurrence of paranoid schizophrenia.

In summary, the following conclusions can be made:

The psychoses of the schizophrenic spectrum do not represent a disease continuum. According to Leonhard it is possible to differentiate between psychoses with a minor hereditary pattern (cycloid psychoses and systematic schizophrenia) and those with a high hereditary pattern (unsystematic schizophrenia). Pregnancy and birth complications appear to be highly etiologically significant for cycloid psychoses and systematic schizophrenia. In this case birth complications are probably only epiphenomena of prior complications during pregnancy. The second trimester of pregnancy is an especially critical phase for prenatal brain development. Exogenous, environmental stressors such as infectious diseases, as well as intrauterine deficiencies and hypoxia (e.g. the "twin transfusion syndrome") can apparently disturb this developmental phase to the point that they contribute to the predestination of the affected individual later developing cycloid psychosis or systematic schizophrenia. In the case of the highly genetically determined unsystematic schizophrenia, pregnancy and birth complications can very likely cause not only the early manifestation of the illness, but often miscarriages, stillbirths, or increased infant and child mortality as well. This may possibly be due to the fact that genetically determined defective brain development is often incompatible with sustaining life in the event of additional exogenous stressors being present. Individuals with a genetic disposition to developing cataphasia may be especially at risk.

Summary

A systematic twin study in the region of Lower Franconia investigated the question of whether the spectrum of schizophrenic and schizophrenia-like psychoses is a disease continuum or whether it encompasses various nosological entities.

The twin survey was limited-representative, i.e. all twins who had been hospitalized in a psychiatric clinic were ascertained. An unlimited representative twin survey was not possible since neither a twin register nor a psychosis register is kept in Germany. According to medical records, 77 index-twins of 66 same-sex pairs suffered from psychoses belonging to the schizophrenic spectrum. Among the 66 pairs there were six cases in which one partner was deceased at the time of the investigation; in eight pairs one partner refused to cooperate in the study; and in five pairs the diagnosis of a psychosis of the schizophrenic spectrum was not confirmed during the personal examination of the index-twins. Therefore a total of 47 same-sex pairs were assessed (22 female and 25 male pairs).

The zygosity of the pairs was determined by molecular-genetic methods using highly polymorphous microsatellites. A questionnaire was also used at the same time. In 43 pairs it was possible to obtain blood samples from both partners in order to determine zygosity. The results of the questionnaire corresponded to the results of the molecular-genetic investigation in 42 cases of the 43 pairs (98%). For four pairs the zygosity diagnosis was based solely on the questionnaire method and the comparison of physical likeness. Twenty-two pairs were monozygotic, while 25 pairs were dizygotic.

The psychiatric diagnoses were carried out by two experienced psychiatric specialists. One investigator diagnosed all the index-twins with no information about the co-twin. The co-twin was diagnosed by a second investigator. Both investigators had no information of the zygosity of the twins at the time of the diagnoses. The diagnostics were based on the operational classification systems of DSM-III-R and ICD 10 as well as Leonhard's nosology. The complete family histories of all test subjects were recorded. If other living family members were mentally ill, they were examined in person. Any existing medical documents of ill family members were also obtained.

An extensive retrospective birth history was recorded for every test subject. Complete information was obtained for all but three of the pairs. In addition, the pre-psychotic role assignment within the pair (dominant/subordinate) and handedness were also established.

It is difficult to equate diagnoses using operational classification systems with diagnoses within the Leonhard classification. Without additional personal data or examination, only the diagnoses "schizophreniform disorder" (DSM-III-R) and "acute transient psychotic disorder" (ICD 10) permit the diagnosis of cycloid psychosis according to Leonhard's definition. The DSM-III-R/ICD 10 category "schizophrenia", however, encompasses cycloid psychoses, unsystematic and systematic schizophrenia. The diagnosis "schizoaffective disorder" also proved to be difficult to match with a Leonhard diagnosis. This term included cycloid psychoses and unsystematic schizophrenia to an almost equal degree.

Following diagnostic differentiation, pre-psychotic irregularities such as poor social integration were found mostly among the periodically catatonic and systematic schizophrenic test subjects when comparing demographic data on the test subjects. All systematically schizophrenic and all periodically catatonic subjects had unfavorable histories of social integration. Test subjects with affective paraphrenia were often well integrated socially (36%) at the time of the follow-up examination. Among test subjects with a cycloid psychoses, this was more often the case, at 67%.

Predominantly concordant pairs were found among unsystematic schizophrenics. In their families there were frequently other ill relatives of the first and second degrees with psychoses of the schizophrenic spectrum, or suicides which occurred without previous psychiatric diagnoses. In contrast, pairs with cycloid psychotic index-twins were predominantly discordant (monozygotic and dizygotic pairs) and 6.6% of relatives of the first degree suffered from affective psychoses. According to Galton's law, unsystematic schizophrenia is determined primarily on a genetic basis. Among the cycloid psychoses, however, genetic disposition was almost irrelevant. The cycloid psychoses indicate by the familial histories of exclusively affective psychoses that they are much closer to the purely affective psychoses than schizophrenia. The exact relationship is still unclear. Twin and family investigations

(Bertelsen et al. 1977) which show that manic-depressive illness reveals a very high genetic pattern, completely in contrast to cycloid psychosis, argue against the assumption that cycloid psychosis is merely a version of manic-depressive illness.

Monozygotic twins were not among those diagnosed with systematic schizophrenia, while this illness occurred among the dizygotic twins at a statistically expected rate. The evaluation of this remarkable finding is complicated and presents a great challenge for further research. Leonhard's hypothesis that a close interpersonal relationship, which often occurs between monozygotic twins, can prevent these severe illnesses, is bold but not implausible, and therefore merits urgent investigation. The hypothesis that a deficient development of ego boundaries ("confusion of ego identity") is the predestination to schizophrenic psychoses may be viewed as refuted, at least in the case of systematic schizophrenia.

Complications during birth play an important etiological role in cycloid psychoses as well as in systematic schizophrenia. In these cases birth complications are probably only epiphenomena of prior pregnancy complications. Apparently, environmental stressors present in certain periods of the extremely critical brain development phase during the second trimester can contribute towards the predestination of an individual to develop cycloid psychosis or systematic schizophrenia later. In the case of unsystematic schizophrenia, birth complications do not contribute greatly to the origin of the illness. However, they can apparently cause a premature and/or more severe course of the illness and are perhaps responsible for the increased rate of miscarriages, stillbirths and increased mortality of infants/children reported among the children of schizophrenics. Individuals with the genetic disposition to cataphasia appear especially at risk. In general, individuals who are genetically disposed to unsystematic schizophrenia and have poor intrauterine conditions as well, which is often the case in twin pregnancies, could possibly be at existential risk in the pre- and/or perinatal stages.

Conclusion

The spectrum of psychoses with schizophrenic and schizophrenia-like symptoms is not a continuum of diseases.

It is possible to define three major groups: cycloid psychoses and unsystematic and systematic schizophrenias. In each of these groups genetic, somatic and psycho-social factors play a completely different etiological role. Therefore, contradictory results are inevitable in the search for common causes of illnesses in the schizophrenic spectrum. Only through a clinical-psychopathological differentiation of individual symptoms and separate and specific scientific research into these illnesses can we expect to make new discoveries in the study of "schizophrenic" psychoses.

Appendix: Synoptic tables

Table A1. Determination of zygosity of the 47 same sex pairs (*1* molecular-genetic method *2* zygosity questionnaire developed by Torgersen 1979)

Pair no.	Zygosity	Probability	Methods
W – 1	monozygotic	99.97%	1 + 2
M – 2	monozygotic	97%	2
M – 3	dizygotic	100%	1 + 2
W – 4	dizygotic	100%	1 + 2
M – 5	dizygotic	100%	1 + 2
M – 6	dizygotic	100%	1 + 2
M – 7	monozygotic	99.99%	1 + 2
M – 8	monozygotic	99.99%	1 + 2
W – 9	monozygotic	99.99%	1 + 2
M – 10	dizygotic	100%	1 + 2
M – 11	monozygotic	99.97%	1 + 2
M – 12	dizygotic	97%	2
M – 13	dizygotic	100%	1 + 2
M – 14	monozygotic	99.99%	1 + 2
M – 15	monozygotic	99.99%	1 + 2
M – 16	monozygotic	99.99%	1 + 2
W – 17	monozygotic	99.99%	1 + 2
M – 18	dizygotic	100%	1 + 2
M – 19	monozygotic	97%	2
W – 20	dizygotic	100%	1 + 2
W – 21	dizygotic	100%	1 + 2
W – 22	dizygotic	100%	1 + 2
W – 23	dizygotic	100%	1 + 2
W – 24	dizygotic	100%	1 + 2
M – 25	monozygotic	99.99%	1 + 2
W – 26	dizygotic	100%	1 + 2
W – 27	monozygotic	99.99%	1 + 2
M – 28	monozygotic	99.99%	1 + 2
W – 29	monozygotic	99.99%	1 + 2
M – 30	dizygotic	100%	1 + 2
W – 31	monozygotic	99.99%	1 + 2
W – 32	monozygotic	99.99%	1 + 2
W – 33	dizygotic	97%	2
M – 34	dizygotic	100%	1 + 2
W – 35	monozygotic	99.97%	1 + 2
W – 36	monozygotic	99.97%	1 + 2
M – 37	monozygotic	99.99%	1 + 2
W – 38	monozygotic	99.97%	1 + 2

Table A1. (continued)

M – 39	dizygotic	100%	1 + 2
M – 40	dizygotic	100%	1 + 2
M – 41	dizygotic	100%	1 + 2
M – 42	monozygotic	99.99%	1 + 2
W – 43	dizygotic	100%	1 + 2
W – 44	dizygotic	100%	1 + 2
W – 45	dizygotic	100%	1 + 2
M – 46	dizygotic	100%	1 + 2
M – 47	dizygotic	100%	1 + 2

Table A2. Diagnoses of the monozygotic subjects using operationalized classifications (*DSM-III-R/ICD 10*)

Pair no.	Index/co	DSM-III-R	ICD 10
W 1–1	index	delusional disorder	delusional disorder
W 1–2	index	delusional disorder	delusional disorder
M 2–1	index	schizoaffective disorder	acute transient psychotic disorder
M 2–2	index	schizoaffective disorder	acute transient psychotic disorder
M 7–1	co	healthy	healthy
M 7–2	index	schizophreniform disorder	acute transient psychotic disorder
M 8–1	index	schizophrenia	schizophrenia
M 8–2	index	schizophrenia	schizophrenia
W 9–1	index	schizophrenia	schizophrenia
W 9–2	index	schizophrenia	schizophrenia
M 11–1	co	healthy	healthy
M 11–2	index	schizophreniform disorder	acute transient psychotic disorder
M 14–1	index	schizoaffective disorder	schizoaffective disorder
M 14–2	co	healthy	healthy
M 15–1	index	schizophrenia	schizophrenia
M 15–2	index	schizophrenia	schizophrenia
M 16–1	co	healthy	healthy
M 16–2	index	schizoaffective disorder	schizoaffective disorder
W 17–1	co	healthy	healthy
W 17–2	index	atypical psychosis	non-organic disorder NOS
M 19–1	co	schizotype personality	schizotype personality
M 19–2	index	schizoaffective disorder	schizoaffective disorder
M 25–1	index	schizophrenia	schizophrenia
M 25–2	co	chronic insomnia	insomnia
W 27–1	index	schizophrenia	schizophrenia
W 27–2	co	schizophreniform disorder	acute transient psychotic disorder
W 28–1	co	healthy	healthy
W 28–2	index	schizophreniform disorder	acute transient psychotic disorder
W 29–1	index	schizophrenia	schizophrenia
W 29–2	index	schizophrenia	schizophrenia
W 31–1	co	healthy	healthy
W 31–2	index	schizoaffective disorder	schizoaffective disorder
W 32–1	co	schizophrenia	schizophrenia
W 32–2	index	schizophrenia	schizophrenia
W 35–1	index	schizophreniform disorder	acute transient psychotic disorder
W 35–2	co	healthy	healthy
W 36–1	index	schizoaffective disorder	schizoaffective disorder
W 36–2	index	schizoaffective disorder	schizoaffective disorder
M 37–1	co	personality disorder NOS	personality disorder NOS
M 37–2	index	schizophrenia	schizophrenia
W 38–1	index	atypical psychosis	other non-organic disorder
W 38–2	index	schizoaffective disorder	schizoaffective disorder
M 42–1	index	schizophrenia	schizophrenia
M 42–2	index	schizophrenia	schizophrenia

Table A3. Diagnoses of the dizygotic subjects using operationalized classifications (*DSM-III-R/ICD 10*)

Pair no.	Index/co	DSM-III-R	ICD 10
M 3–1	co	healthy	healthy
M 3–2	index	schizophrenia	schizophrenia
W 4–1	index	delusional disorder	delusional disorder
W 4–2	co	healthy	healthy
M 5–1	index	schizophrenia	schizophrenia
M 5–2	co	healthy	healthy
M 6–1	co	avoidant personality	anxious-avoidant personality
M 6–2	index	schizophrenia	schizophrenia
W 10–1	index	delusional disorder	delusional disorder
W 10–2	co	borderline disorder	emotionally unstable personality
M 12–1	index	schizophrenia	schizophrenia
M 12–2	co	schizophrenia	schizophrenia
M 13–1	index	schizophrenia	schizophrenia
M 13–2	co	healthy	healthy
M 18–1	co	healthy	healthy
M 18–2	index	schizophreniform disorder	acute transient psychotic disorder
W 20–1	co	healthy	healthy
W 20–2	index	schizophrenia	schizophrenia
W 21–1	index	schizoaffective disorder	schizoaffective disorder
W 21–2	index	schizoaffective disorder	schizoaffective disorder
W 22–1	co	healthy	healthy
W 22–2	index	atypical psychosis	non-organic disorder NOS
W 23–1	index	schizoaffective disorder	schizoaffective disorder
W 23–2	co	healthy	healthy
W 24–1	co	healthy	healthy
W 24–2	index	schizoaffective disorder	schizoaffective disorder
W 26–1/2	co	healthy	healthy
W 26–3	index	atypical psychosis	acute transient psychotic disorder
M 30–1	co	healthy	healthy
M 30–2	index	schizophrenia	schizophrenia
M 33–1	co	healthy	healthy
M 33–2	index	schizophrenia	schizophrenia
M 34–1	index	schizophrenia	schizophrenia
M 34–2	co	healthy	healthy
M 39–1	index	schizophrenia	schizophrenia
M 39–2	co	healthy	healthy
M 40–1	index	schizoaffective disorder	schizoaffective disorder
M 40–2	index	schizoaffective disorder	schizoaffective disorder
M 41–1	index	schizophrenia	schizophrenia
M 41–2	co	schizophreniform disorder	acute transient psychotic disorder
W 43–1	co	schizophrenia	schizophrenia
W 43–2	index	schizophrenia	schizophrenia
W 44–1	index	schizoaffective disorder	schizoaffective disorder
W 44–2	co	healthy	healthy

Table A3. (continued)

W 45–1	co	healthy	healthy
W 45–2	index	delusional disorder	delusional disorder
M 46–1	co	healthy	healthy
M 46–2	index	schizoaffective disorder	schizoaffective disorder
M 47–1	co	healthy	healthy
M 47–2	index	schizophreniform disorder	acute transient psychotic disorder

Table A4. Diagnoses of the monozygotic subjects using the *Leonhard Classification*

Pair no.	Index/co	*Leonhard diagnosis*
W 1–1	index	affective paraphrenia
W 1–2	index	affective paraphrenia
M 2–1	index	motility psychosis
M 2–2	index	motility psychosis
M 7–1	co	healthy
M 7–2	index	anxiety psychosis
M 8–1	index	periodic catatonia
M 8–2	index	periodic catatonia
W 9–1	index	periodic catatonia
W 9–2	index	periodic catatonia
M 11–1	co	healthy
M 11–2	index	anxiety-happiness psychosis
M 14–1	index	confusion psychosis
M 14–2	co	healthy
M 15–1	index	periodic catatonia
M 15–2	index	periodic catatonia
M 16–1	co	healthy
M 16–2	index	anxiety psychosis
W 17–1	co	healthy
W 17–2	index	anxiety psychosis
M 19–1	co	periodic catatonia
M 19–2	index	periodic catatonia
M 25–1	index	affective paraphrenia
M 25–2	co	chronic insomnia
W 27–1	index	motility psychosis
W 27–2	co	motility psychosis
M 28–1	co	healthy
M 28–2	index	anxiety psychosis
W 29–1	index	motility psychosis
W 29–2	index	motility psychosis
W 31–1	co	healthy
W 31–2	index	anxiety psychosis
W 32–1	co	periodic catatonia
W 32–2	index	periodic catatonia
W 35–1	index	confusion psychosis
W 35–2	co	healthy
W 36–1	index	affective paraphrenia
W 36–2	index	affective paraphrenia
M 37–1	co	abnormal personality
M 37–2	index	periodic catatonia
W 38–1	index	affective paraphrenia
W 38–2	index	affective paraphrenia
M 42–1	index	cataphasia
M 42–2	index	cataphasia

Table A5. Diagnoses of the dizygotic subjects using the *Leonhard Classification*

Pair no.	Index/co	Leonhard diagnosis
M 3–1	co	healthy
M 3–2	index	systematic catatonia
W 4–1	index	affective paraphrenia
W 4–2	co	healthy
M 5–1	index	hebephrenia
M 5–2	co	healthy
M 6–1	co	emotionally unstable personality
M 6–2	index	affective paraphrenia
W 10–1	index	affective paraphrenia
W 10–2	co	abnormal personality
M 12–1	index	periodic catatonia
M 12–2	co	periodic catatonia
M 13–1	index	systematic catatonia
M 13–2	co	healthy
M 18–1	co	healthy
M 18–2	index	anxiety psychosis
W 20–1	co	healthy
W 20–2	index	hebephrenia
W 21–1	index	anxiety psychosis
W 21–2	index	anxiety psychosis
W 22–1	co	healthy
W 22–2	index	periodic catatonia
W 23–1	index	anxiety psychosis
W 23–2	co	healthy
W 24–1	co	healthy
W 24–2	index	anxiety-happiness psychosis
W 26–1/2	co	healthy
W 26–3	index	confusion psychosis
M 30–1	co	healthy
M 30–2	index	hebephrenia
W 33–1	co	healthy
W 33–2	index	motility psychosis
M 34–1	index	affective paraphrenia
M 34–2	co	healthy
M 39–1	index	systematic paraphrenia
M 39–2	co	healthy
M 40–1	index	confusion psychosis
M 40–2	index	confusion psychosis
M 41–1	index	affective paraphrenia
M 41–2	co	affective paraphrenia (mild residual state)
W 43–1	co	abnormal personality
W 43–2	index	periodic catatonia
W 44–1	index	anxiety-happiness psychosis
W 44–2	co	healthy

Table A5. (continued)

W 45–1	co	healthy
W 45–2	index	anxiety-happiness psychosis
M 46–1	co	healthy
M 46–2	index	anxiety-happiness psychosis
M 47–1	co	healthy
M 47–2	index	anxiety psychosis

Table A6. Education, social situation and marital status (at first hospitalization and follow-up study) of the monozygotic subjects

Pair no.	Education	Social situation 1st hosp./follow-up.	Marital status 1st hosp./follow-up.	
W 1–1	higher-level school	employed/homeless	single	single
W 1–2	higher-level school	employed/homeless	single	single
M 2–1	middle school	employed/employed	single	single
M 2–2	middle school	employed/unemployed	single	single
M 7–1	higher-level school	employed		married
M 7–2	middle school	employed/unemployed	single	single
M 8–1	no school completed	nursing home/permanently hospitalized	single	single
M 8–2	no school completed	nursing home/permanently hospitalized	single	single
W 9–1	school for learning disabled children	home care/permanently hospitalized	single	single
W 9–2	school for learning disabled children	home care/home for the elderly	single	single
M 11–1	higher-level school	student		single
M 11–2	higher-level school	student/student	single	single
M 14–1	middle school	employed/unemployed	single	single
M 14–2	middle school	armed forces		single
M 15–1	school for learning disabled children	employed/permanently hospitalized	single	single
M 15–2	school for learning disabled children	unemployed/permanently hospitalized	single	single
M 16–1	middle school	employed		married
M 16–2	middle school	employed/employed	single	married
W 17–1	school for learning disabled children	homemaker		married
W 17–2	school for learning disabled children	employed/home for the disabled	single	single
M 19–1	middle school	employed/unemployed	single	single
M 19–2	middle school	unemployed/rehabilitation	single	single
M 25–1	middle school	employed/early retirement	single	single
M 25–2	middle school	employed		married
W 27–1	higher-level school	employed/retirement	single	single
W 27–2	higher-level school	employed/retirement	single	single

Table A6. (continued)

M 28–1	middle school	employed		married
M 28–2	middle school	employed/employed	single	single
W 29–1	higher-level school	apprentice/early retirement	single	single
W 29–2	higher-level school	apprentice/early retirement	single	single
W 31–1	higher-level school	employed		married
W 31–2	middle school	employed/homemaker	single	married
W 32–1	middle school	employed/permanently hospitalized	single	single
W 32–1	middle school	employed/permanently hospitalized	single	single
W 35–1	middle school	employed/homemaker	married	married
W 35–2	middle school	homemaker		married
W 36–1	middle school	homemaker/homemaker	married	married
W 36–2	middle school	pupil/homemaker	single	married
M 37–1	higher-level school	employed		married
M 37–2	school for learning disabled children	apprentice/permanently hospitalized	single	single
W 38–1	higher-level school	student/unemployed	single	single
W 38–2	higher-level school	employed/unemployed	single	single
M 42–1	school for learning disabled children	employed/temporary retirement	single	single
M 42–2	school for learning disabled children	employed/unemployed	single	single

Table A7. Education, social situation and marital status (at first hospitalization and follow-up study) of the dizygotic subjects

Pair no.	Education	Social situation 1st hosp./follow-up.	Marital status 1st hosp./follow-up.	
M 3–1	higher-level school	employed		married
M 3–2	school for learning disabled children	unemployed/permanently hospitalized	single	single
W 4–1	middle school	employed/employed	married	married
W 4–2	middle school	homemaker		married
M 5–1	higher-level school	unemployed/unemployed	single	single
M 5–2	higher-level school	employed		married
M 6–1	higher-level school	employed		married
M 6–2	higher-level school	employed/early retirement	married	divorced
W 10–1	higher-level school	employed/unemployed	single	divorced
W 10–2	higher-level school	employed		married
M 12–1	school for learning disabled children	employed/unemployed	single	single
M 12–2	school for learning disabled children	unemployed/homeless	single	single
M 13–1	no school completed	home care/nursing home	single	single
M 13–2	middle school	employed		divorced
M 18–1	higher-level school	student		single
M 18–2	higher-level school	pupil/employed	single	single
W 20–1	middle school	employed		single
W 20–2	middle school	unemployed/group home	single	single
W 21–1	middle school	homemaker/homemaker	married	married
W 21–2	middle school	homemaker/homemaker	married	married
W 22–1	middle school	homemaker		single[1]
W 22–2	middle school	pupil/unemployed	single	single
W 23–1	higher-level school	employed/employed	married	divorced
W 23–2	higher-level school	employed		single[1]
W 24–1	school for learning disabled children	employed		married
W 24–2	school for learning disabled children	homemaker/hospitalized	married	married
W 26–1/2	middle school	employed		married
W 26–3	middle school	employed/early retirement[3]	single	single
M 30–1	higher-level school	student		married
M 30–2	higher-level school	pupil/group home	single	single

Table A7. (continued)

W 33–1	middle school	homemaker		married
W 33–2	middle school	employed/early retirement	single	widowed
M 34–1	higher-level school	student/unemployed	single	single
M 34–2	higher-level school	employed		single[1]
M 39–1	higher-level school	unemployed/unemployed	single	single
M 39–2	higher-level school	employed		single
M 40–1	middle school	unemployed/unemployed	single	single
M 40–2	middle school	unemployed/unemployed	single	married
M 41–1	higher-level school	apprentice/unemployed	single	single
M 41–2	higher-level school	pupil /employed	single	married
W43–1	school for learning disabled children	employed/employed	single	single[1]
W43–2	school for learning disabled children	employed/social welfare	single	single
W 44–1	school for learning disabled children	employed/unemployed	single[1]	single
W 44–2	school for learning disabled children	employed		married
W 45–1	middle school	homemaker		married
W 45–2	middle school	employed/early retirement	single	single
M 46–1	higher-level school	employed		married
M 46–2	higher-level school	student/odd jobs	single	single[1]
M 47–1	middle school	employed		married
M 47–2	middle school	employed/employed	single	single

[1] Lives together with long-term partner.
[2] Took care of the household until hospitalization.
[3] After the first psychotic phase, suffered a severe accident with a polytrauma.

Table A8. Table of retrospectively obtained data of the birth histories of the monozygotic subjects (Fuchs-rating-scale of complications during pregnancy and birth from *Parnas et al.* 1982 was used to determine the number and severity of complications). The ill or more severely ill subject is always referred to first

Pair no.	Birth order	Birth weight	Birth complications number	severity
W 1	2 – 1	3000g – 2500g	0 – 1	0 – 2
M 2[1]	2 – 1	2800g – 2800g	2 – 0	5 – 0
M 7	2 – 1	2550g – 2700g	2 – 1	5 – 1
M 8[2]	1 – 2	2500g – 2100g	2 – 3	3 – 10
M 9[2]	1 – 2	3200g – 3000g	1 – 1	4 – 4
M 11	2 – 1	2500g – 2500g	3 – 0	7 – 0
M 14	1 – 2	3000g – 3100g	1 – 0	1 – 0
M 15	2 – 1	2800g – 3000g	1 – 1	2 – 2
M 16	2 – 1	2500g – 3000g	1 – 0	2 – 0
W 17	2 – 1	2400g – 2500g	0 – 0	0 – 0
M 19	2 – 1	2100g – 2400g	1 – 2	4 – 5
M 25	1 – 2	2000g – 2300g	1 – 2	4 – 5
W 27	1 – 2	1700g – 2200g	1 – 1	4 – 4
M 28	2 – 1	2800g – 2800g	1 – 0	2 – 0
W 29	2 – 1	1800g – 1300g	1 – 1	4 – 4
W 31	2 – 1	2900g – 3000g	0 – 0	0 – 0
W 32	2 – 1	1600g – 1600g	1 – 1	4 – 4
W 35	1 – 2	2600g – 2100g	1 – 0	4 – 0
W 36	2 – 1	2600g – 2100g	0 – 0	0 – 0
W 37	2 – 1	1700g – 2800g	6 – 3	16 – 6
W 38	2 – 1	2200g – 2400g	3 – 1	6 – 1
M 42[2]	1 – 2	3250g – 3250g	0 – 1	0 – 4

[1] First and second born of triplets. The other triplet is healthy, weighed 2500g and had no birth complications.
[2] There was little difference in the severity of illness of both subjects.

Table A9. Table of retrospectively obtained data of the birth histories of the dizygotic subjects (Fuchs-rating-scale of complications during pregnancy and birth from *Parnas et al.* 1982 was used to determine the number and severity of complications). The ill or more severely ill subject is always referred to first

Pair no.	Birth order	Birth weight	Birth complications number	severity
M 3	2 – 1	2800g – 2500g	5 – 0	11 – 0
W 4	1 – 2	1800g – 1500g	3 – 3	9 – 9
M 5	1 – 2	1500g – 1500g	2 – 2	6 – 6
M 6	2 – 1	3700g – 3500g	4 – 2	9 – 3
W 10	1 – 2	2500g – 2000g	1 – 1	2 – 4
M 12	1 – 2	3200g – 3400g	0 – 0	0 – 0
M 13	1 – 2	2500g – 2300g	0 – 0	0 – 0
M 18	2 – 1	2700g – 3000g	0 – 0	0 – 0
W 20	2 – 1	2500g – 2300g	2 – 0	6 – 0
W 21[1]	2 – 3	1800g – 1300g	1 – 1	4 – 4
W 22	2 – 1	unknown	1 – 0	2 – 0
W 23	1 – 2	1300g – 1700g	1 – 1	4 – 4
W 24	2 – 1	1900g – 2100g	3 – 3	9 – 9
W 26[2]	3 – 1/2	1800g – 2000g	1 – 2/2	
M 30	2 – 1	3000g – 3000g	2 – 0	6 – 0
W 33	2 – 1	unknown	unknown	
M 34	1 – 2	2000g – 2500g	0 – 0	0 – 0
M 39	1 – 2	2500g – 1700g	2 – 2	3 – 5
M 40	1 – 2	1500g – 1500g	1 – 1	3 – 3
M 41	1 – 2	3350g – 3250g	0 – 0	0 – 0
W 43	1 – 2	4000g – 4000g	0 – 0	0 – 0
W 44	1 – 2	2000g – 3700g	3 – 2	9 – 7
W 45	2 – 1	unknown	unknown	
M 46	2 – 1	3150g – 3300g	3 – 2	7 – 2
M 47	2 – 1	4500g – 3500g	1 – 1	3 – 3

[1] The first born triplet was deceased. She had always been mentally healthy, had weighed 2300 g at birth and had had no birth complications.

[2] The subject is the third born of triplets. Her two other sisters are healthy.

Table A10. Handedness of monozygotic subjects (*RH* strictly right-handed; *NRH* non-right-handed, where the ambidextrous and strictly left-handers were combined)

Pair no.	Index/co	Concordant C discordant D	Handedness
W 1 – 1	index		RH
		C1	
W 1 – 2	index		RH
M 2 – 1[1]	index		RH
		C1	
M 2 – 2	index		RH
M 7 – 1	co		*NRH*
		D	
M 7 – 2	index		RH
M 8 – 1	index		RH
		C1	
M 8 – 2	index		RH
W 9 – 1	index		RH
		C1	
W 9 – 2	index		RH
M 11 – 1	co		RH
		D	
M 11 – 2	index		RH
M 14 – 1	index		RH
		C3	
M 14 – 2	co		*NRH*
M 15 – 1	index		RH
		C1	
M 15 – 2	index		RH
M 16 – 1	co		RH
		D	
M 16 – 2	index		RH
W 17 – 1	co		RH
		D	
W 17 – 2	index		RH
M 19 – 1	co		RH
		C1	
M 19 – 2	index		RH
M 25 – 1	index		*NRH*
		C3	
M 25 – 2	co		RH
W 27 – 1	index		RH
		C1	
W 27 – 2	co		RH

Table A10. (continued)

M 28 – 1	co		RH
		D	
M 28 – 2	index		RH
W 29 – 1	index		*NRH*
		C1	
W 29 – 2	index		RH
W 31 – 1	co		RH
		D	
W 31 – 2	index		RH
W 32 – 1	co		RH
		C1	
W 32 – 2	index		RH
W 35 – 1	index		RH
		D	
W 35 – 2	co		*NRH*
W 36 – 1	index		RH
		C1	
W 36 – 2	index		RH
M 37 – 1	co		RH
		C3	
M 37 – 2	index		RH
W 38 – 1	index		RH
		C1	
W 38 – 2	index		RH
M 42 – 1	index		RH
		C1	
M 42 – 2	index		RH

[1] The subjects have a third born triplet sister (*RH*).

Table A11. Handedness of dizygotic subjects (*RH* strictly right-handed; *NRH* non-right-handed, where the ambidextrous and strictly left-handers were combined)

Pair no.	Index/co	Concordant C discordant D	Handedness
M 3 – 1	co		*NRH*
		D	
M 3 – 2	index		RH
W 4 – 1	index		RH
		D	
W 4 – 2	co		RH
M 5 – 1	index		RH
		D	
M 5 – 2	co		RH
M 6 – 1	co		RH
		C3	
M 6 – 2	index		*NRH*
W 10 – 1	index		RH
		C3	
W 10 – 2	co		RH
M 12 – 1	index		RH
		C1	
M 12 – 2	co		RH
M 13 – 1	index		RH
		D	
M 13 – 2	co		*NRH*
M 18 – 1	co		RH
		D	
M 18 – 2	index		*NRH*
W 20 – 1	co		RH
		D	
W 20 – 2	index		RH
W 21 – 1	index		RH
		C1	
W 21 – 2	index		RH
W 22 – 1	co		RH
		D	
W 22 – 2	index		RH
W 23 – 1	index		*NRH*
		D	
W 23 – 2	co		*NRH*
W 24 – 1	co		RH
		D	
W 24 – 2	index		RH

Table A11. (continued)

W 26 – 1/2[1]	co		RH
		D	
W 26 – 3	index		RH
M 30 – 1	co		RH
		D	
M 30 – 2	index		RH
W 33 – 1	co		RH
		D	
W 33 – 2	index		RH
M 34 – 1	index		RH
		D	
M 34 – 2	co		RH
M 39 – 1	index		RH
		D	
M 39 – 2	co		RH
M 40 – 1	index		RH
		C1	
M 40 – 2	index		RH
M 41 – 1	index		RH
		C2	
M 41 – 2	co		RH
W 43 – 1	co		RH
		C1	
W 43 – 2	index		RH
W 44 – 1	index		RH
		D	
W 44 – 2	co		RH
W 45 – 1	co		RH
		D	
W 45 – 2	index		RH
M 46 – 1	co		RH
		D	
M 46 – 2	index		*NRH*
M 47 – 1	co		*NRH*
		D	
M 47 – 2	index		*NRH*

[1] W 26 – 1/2 are the healthy first and second born of triplet sisters (all are *RH*).

Bibliography

Achs R, Harper RG, Siegel M (1966) Unusual dermatoglyphic findings associated with rubella embryopathy. N Engl J Med 274:148–150.

Ackerknecht EH (1985) Kurze Geschichte der Psychiatrie. Enke, Stuttgart.

Adams W, Kendell RE, Hare EH, Munk-Jörgensen P (1993) Epidemiological evidence that maternal influenza contributes to the aetiology of schizophrenia. Br J Psychiatry 163:522–534.

Akbarian S, Bunney WEjr, Potkin SG, Wigal SB, Habman JO, Sandman CA, Jones EG (1993) Altered distribution of nicotin-amide-adenine dinucleotide phosphate-diaphorase cells in frontal lobe of schizophrenics implies disturbance of cortical development. Arch Gen Psychiatry 50:169–177.

Allen G (1979) Holzinger's Hc revised. Acta Genet Med Gemellol 28:161–164.

Allen G, Hrubec Z (1979) Twin concordance: a more general model. Acta Genet Med Gemellol 28:3–13.

Allen G, Parisi GA (1990) Trends in monozygotic and dizygotic twinning rates by maternal age and parity. Acta Genet Med Gemellol 30:317–328.

Allen G, Pollin W (1970) Schizophrenia in twins and the diffuse ego boundary hypothesis. Am J Psychiatry 127:437–442.

Allen G, Harvald B, Shields J (1967) Measures of twin concordance. Acta Gen et Stat Med 17:475–481.

Alter M, Schulenberg R (1966) Dermatoglyphics in the rubella syndrome. JAMA 197:685–688.

American Psychiatric Association (1987) Diagnostic and statistic manual of mental disorders, 3rd ed, revised. APA, Washington DC.

American Psychiatric Association (1994) Diagnostic and statistic manual of mental disorders. 4th ed. APA, Washington DC.

Andreasen NC, Olsen SA (1982) Negative vs positive schizophrenia. Arch Gen Psychiatry 39:789–794.

Annett M (1970) A classification of hand preference by association analyses. Br J Psychology 61:303–321.

Arnold SE, Hyman BT, van Hoesen GW (1991) Cytoarchitectural abnormalities of the entorhinal cortex in schizophrenia. Arch Gen Psychiatry 48:625–632.

Baron M, Levitt M (1980) Platelet monoamine oxidase activity: relation to genetic load of schizophrenia. Psychiatry Res 3:69–74.

Barr A, Stevenson AC (1961) Stillbirths and infant mortality in twins. Ann Hum Genet 25:131–140.

Barr CE, Mednick SA, Munk-Jorgensen P (1990) Exposure to influenza epidemics during gestation and adult schizophrenia. Arch Gen Psychiatry 47:869–874.

Bartley AJ, Jones DW, Weinberger DR (1997) Genetic variability of human brain size and cortical gyral patterns. Brain 120 (Part 2):257–269.

134   Bibliography

Bateson G, Jackson D, Haley J (1956) Toward a theory of schizophrenia. Behav Sci
     1:251–264.
Beckmann H, Franzek E (1992) Deficit of birthrates in winter and spring months in dis-
     tinct subgroups of mainly genetically determined schizophrenia. Psychopathology
     25:57–64.
Beckmann H, Franzek E (1992) The influence of neuroleptics on specific syndromes and
     symptoms in schizophrenics with unfavourable long-term course. Neuropsychobiology
     (Pharmacopsychiatry) 26:50–58.
Beckmann H, Jakob H (1991) Prenatal disturbances of nerve cell migration in the
     entorhinal region: a common vulnerability factor in functional psychoses ? J Neural
     Transm (GenSect) 84:155–164.
Beckmann H, Fritze J, Lanczik M (1990) Prognostic validity of the cycloid psychoses. A
     prospective follow-up study. Psychopathology 23:205–211.
Beckmann H, Franzek E, Stöber G (1996) Genetic heterogeneity in catatonic schizophre-
     nia: a family study. Am J Med Gen (Neuropsychiatric Sec) 67:289–300.
Belmaker RH, Pollin W, Wyatt RJ, Cohen S (1974) A follow-up of monozygotic twins
     discordant for schizophrenia. Arch Gen Psychiatry 30:219–222.
Benirschke K, Kim CK (1973) Multiple pregnancy. N Engl J Med 288:1276–1284.
Bergström K, Bille B, Rasmussen F (1984) Computed tomography of the brain in child-
     ren with minor neurodevelopmental disorders. Neuropediatrics 15:115–119.
Bertelsen A, Harvald B, Hauge M (1977) A Danish twin study of manic-depressive dis-
     orders. Br J Psychiatry 130:330–351.
Bischoff A (1959) Zur Pathologie des sozialen Paarverhaltens von Zwillingen in der
     Entwicklung. Diss., Berlin.
Bleuler E (1911) Dementia praecox oder Gruppe der Schizophrenien. Barth, Leipzig Wien.
Boklage CE (1977) Schizophrenia, brain asymmetry development, and twinning: Cellu-
     lar relationship with etiological and possibly prognostic implications. Biol Psychiatry
     12:19–35.
Braha HS, Torrey EF, Bigelow LB, Lohr JB, Linington BB (1991) Subtle signs of prena-
     tal maldevelopment of the hand ectoderm in schizophrenia: a preliminary monozygotic
     twin study. Biol Psychiatry 30:719–725.
Bracha HS, Torrey EF, Gottesman II, Bigelow LB, Cunniff C (1992) Second trimester
     markers of fetal size in schizophrenia: a study of monozygotic twins. Am J Psychiatry
     149:1355–1361.
Bracken H (1969) Humangenetische Psychologie. In: Becker PE (Hrsg.) Humangene-
     tik, ein kurzes Handbuch, Vol I/2. Thieme, Stuttgart, pp 409–562.
Bradbury TN, Miller GA (1985) Season of birth in schizophrenia: a review of evidence,
     methodology, and etiology. Psychol Bull 98:569–594.
Breslin NA, Weinberger DR (1991) Neurodevelopmental implications of findings from
     brain imagine studies of schizophrenia. In: Mednick SA, Cannon TD, Barr CE, Lyon
     M (eds.) Fetal neural development and adult schizophrenia. Cambridge University
     Press, Cambridge New York Port Chester Melbourne Syndney, pp 199–215.
Brockington IF, Perris C, Kendell RE, Hillier VE, Wainwright S (1982) The course and
     outcome of cycloid psychosis. Psychol Med 12:97–105.
Buchsbaum MS, Mirsky AF, DeLisi LE, Morhisha J, Karson CN, Mendelson WB,
     King AC, Johnson J, Kessler R (1984) The genain quadruplets: elektrophysiological,
     positron emission, and X-ray tomographic studies. Psychiatry Res 13:95–108.
Bulmer MG (1970) The biology of twinning in man. Clarendon Press, Oxford.
Campion E, Tucker G (1973) A note on twin studies, schizophrenia and neurological
     impairment. Arch Gen Psychiatry 29:460–464.
Cannon TD (1991) Genetic and perinatal sources of structural brain abnormalities in
     schizophrenia. In: Mednick SA, Cannon TD, Barr CE, Lyon M (eds.) Fetal neural

development and adult schizophrenia. Cambridge University Press, Cambridge, pp 174–198.

Cannon TD, Mednick SA, Parnas J, Schulsinger F, Praestholm J, Vestergaard A (1993) Developmental brain abnormalities in the offspring of schizophrenic mothers. Arch Gen Psychiatry 50:551–564.

Casanova MF, Sanders RD, Goldberg TE, Bigelow LB, Christison G, Torrey EF, Weinberger DR (1990) Morphometry of the corpus callosum in monozygotic twins discordant for schizophrenia: a magnetic resonance imaging study. J Neurology Neurosurgery Psychiatry 53:416–421.

Cederlöf R, Friberg L, Johannson E, Kaij L (1961) Studies on similarity diagnosis in twins with the aid of mailed questionaires. Acta Genet Statist Med 11:338–362.

D'Amato D, Dalery J, Rochet T, Terra JL, Marie-Cardine M (1991) Saisons de naissance et psychiatrie. Etude retrospective d'une population hospitaliere. Encephale 17:67–71.

Davis J, Phelps A, Bracha HS (1995) Prenatal development of monozygotic twins and concordance for schizophrenia. Schizophr Bull 21:13–18.

Erdmann J, Nöthen M, Stratmann M, Fimmers R, Franzek E, Propping P (1993) The use of microsatellites in zygosity diagnosis of twins. Acta Genet Med Gemellol 42:45–51.

Essen-Möller E (1941) Psychiatrische Untersuchungen an einer Serie von Zwillingen. Acta Psychiat Neurol Scand [Suppl] 23.

Essen-Möller E (1970) Twenty-one psychiatric cases and their co-twins. Acta Genet (Basel) 19:315–317.

Falkai P, Bogerts B, Romuzek M (1988) Limbic pathology in schizophrenia: the entorhinal region – a morphometric study. Biol Psychiatry 24:515–521.

Feinleig M (1985) National Center for Health Statistics. Comparability of reporting between the birth Certificate and the 1980 National Natality Survey. Vital and Health Statistics, Series 2, No 99. DHHS Publication No (PHS) 86–1373. Superindentent of Documents, U.S. Government Printing Office, Washington DC.

Fischer M (1971) Psychoses in the offspring of schizophrenic monozygotic twins and their normal co-twins. Br J Psychiatry 118:43–52.

Fischer M, Harvald B, Hauge M (1969) A Danish twin study of schizophrenia. Br J Psychiatry 115:981–990.

Flor-Henry P (1985) Schizophrenia: sex differences. Can J Psychiatry 30:319–322.

Fogel BJ, Nitowsky HM, Gruenwald P (1965) Discordant abnormalities in monozygotic twins. J Pediatrics 66:64–72.

Franzek E, Beckmann H (1991) Symptom- und Syndromanalyse schizophrener Langzeitverläufe. Nervenarzt 62:549–556.

Franzek E, Beckmann H (1992a) Schizophrenia: not a disease entity? A study of 57 long-term hospitalized chronic schizophrenics. Europ J Psychiatry 6:97–108.

Franzek E, Beckmann H (1992b) Sex differences and distinct subgroups in schizophrenia. Psychopathology 25:90–99.

Franzek E, Beckmann H (1992c) Season-of-birth effect reveals the existence of etiologically different groups of schizophrenia. Biol Psychiatry 32:375–378.

Franzek E, Beckmann H (1996) Gene-Environment interaction in schizophrenia: season-of-birth effect reveals etiologically different subgroups. Psychopathology 29:14–26.

Franzek E, Schmidtke A, Beckmann H, Stöber G (1995) Evidence against unusual sex concordance and pseudoautosomal inheritance in the catatonic subtype of schizophrenia. Psychiatry Res 59:17–24.

Franzek E, Becker T, Hofmann E, Flöhl W, Beckmann H (1996) Is Computerized Tomography ventricular abnormality related to cycloid psychosis? Biol Psychiatry 40:1255–1266.

Franzek E, Beckmann H (1998) Different genetic background of schizophrenia spectrum psychoses: a twin study. Am J Psychiatry 155:76–83.

Friedrich W, Kabat vel Job O (1986) Zwillingsforschung international. Deutscher Verlag der Wissenschaften, Berlin.

Galton F (1876) The history of twins as a criterion of the relative powers of nature and nurture. J Anthropol Inst 5:391–406.

Gayle HD, Yip R, Frank MJ, Nieburg P, Binkin NJ (1988) Validation of maternally reported birth weights among 46637 Tennessee WIC program participants. Publ Health Rep 103:143–146.

Gilmore JH, Perkins DO, Kliewer MA, Hage ML, Silva SG, Chescheir NC, Hertzberg BS, Sears CA (1996) Fetal brain development of twins assessed in utero by ultrasound: implications for schizophrenia. Schizophr Res 19:141–149.

Goedert JJ, Duliege AM, Amos CI, Felton S, Biggar RJ (1991) High-risk of HIV-1 infection for first-born twins. Lancet 338:1471–1475.

Goldberg TE, Ragland JD, Torrey EF, Gold JM, Bigelow LB, Weinberger DR (1990) Neuropsychological assessment of monozygotic twins discordant for schizophrenia. Arch Gen Psychiatry 47:1066–1072.

Goodman R (1988) Are complications of pregnancy and birth causes of schizophrenia? Dev Med Child Neurol 30:391–395.

Gottesman II, Bertelsen A (1989) Confirming unexpressed genotypes for schizophrenia. Risks in the offspring of Fisher's Danish identical and fraternal discordant twins. Arch Gen Psychiatry 46:867–872.

Gottesman JI, Shields J (1966) Schizophrenia in twins: 16 years consecutive admissions to a psychiatric clinic. Br J Psychiatry 112:809–818.

Gottesman II, Shields J (1972) Schizophrenia and genetics: twin study vantage point. Academic Press, New York.

Gottesman II, Shields J (1976) A critical review of recent adoption, twin and family studies of schizophrenia: behavioural genetics perspectives. Schizophr Bull 2:360–401.

Gottesman II, Shields J (1982) Schizophrenia: the epigenetic puzzle. Cambridge University Press, Cambridge.

Gruhle HW (1932) Theorie der Schizophrenie. In: Bumke O (Hrsg.) Handbuch der Geisteskrankheiten. Neunter Band. Spezieller Teil V. Die Schizophrenie. Springer, Berlin Heidelberg New York.

Häfner H, Behrens S, DeVry J, Gattaz WF, Löffler W, Maurer K, Riecher-Rössler A (1991) Warum erkranken Frauen später an Schizophrenie? Erhöhung der Vulnerabilitätsschwelle durch Östrogen. Nervenheilkunde 10:154–163.

Hamilton WJ, Boyd JD, Mossman HW (1972) Human embryology: prenatal development of form and function. pp 567–646. Heffer, Cambridge.

Harrow M, Quinlan D (1977) Is disordered thinking unique to schizophrenia? Arch Gen Psychiatry 34:15–21.

Harvald B, Hauge M (1965) Hereditary factors elucidated by twin studies. In: Neel JV (ed.) Genetics and the epidemiology of chronic diseases. pp 61–76. Department of Health, Education and Welfare. Public Health Service Puplication no 1163.

Heady JA, Heasman MA (1959) Social and biological factors in infant mortality. General register office, studies on medical and population subjects, 159 HM Stationary Office, London.

Heinroth JC (1818) Lehrbuch der Störungen des Seelenlebens oder der Seelenstörungen und ihrer Behandlung. Bd.1. Vogel, Leipzig.

Hoffer A, Pollin W (1970) Schizophrenia in the NAS-NCR panel of 15909 veteran twin pairs. Arch Gen Psychiatry 23:469–477.

Holt SB (1986) The Genetics of dermal ridges. Charles C Thomas, Springfield IL.

Holzinger KJ (1929) The relative effect of nature and nurture on twin differences. J Educ Psychol 20:241–248.

Inouye E (1961) Similarity and dissimilarity of schizophrenia in twins. Proceedings of the Third World Congress on Psychiatry. Vol 1. pp 542–530. University of Toronto Press, Montreal.

Inouye E (1972) Monozygotic twins with schizophrenia reared apart in infancy. Jap J Hum Genet 16:182–190.

Jackson DD (1959) A critique of the literature on the genetics of schizophrenia. In: Jackson DD (ed.) The study of schizophrenia. Basic Books, New York, pp 37–90.

Jakob H, Beckmann H (1984) Clinical-neuropathological studies of developmental disorders in the limbic system in chronic schizophrenia. In: schizophrenia. An Integrative View. XIV Congress CINP. Ricerca Scientifica ed Educazione Permanente. 39: Suppl. 81.

Jakob H, Beckmann H (1986) Prenatal developmental disturbances in the limbic allocortex in schizophrenia. J Neural Transmission 65:303–326.

Johnstone EC, Crow TJ, Frith CD, Husband J, Kreel L (1976) Cerebral ventricular size and cognitive impairment in chronic schizophrenia. The Lancet 2:924–926.

Kabat vel Job O (1986) Verfahren zur Bestimmung der Zygosität. In: Friedrich W, Kabat vel Job O (Hrsg.) Zwillingsforschung international. Verlag der Wissenschaften, Berlin.

Kahlbaum K (1863) Die Gruppierung der psychischen Krankheiten und die Einteilung der Seelenstörungen. Kofemann, Danzig.

Kaiya H, Uematsu M, Ofuji M, Nishida A, Morikiyo M, Adachi S (1989) Computerized tomography in schizophrenia: familial versus non-familial forms of illness. Br J Psychiatry 155:444–450.

Kallmann FJ (1938) The genetics of schizophrenia. Augustin, New York.

Kallmann FJ (1946) The genetic theory of schizophrenia: An analysis of 691 schizophrenic twin index families. Am J Psychiatry 103:309–322.

Kallmann FJ, Roth B (1956) Genetic aspects of preadolescent schizophrenia. Am J Psychiatry 112:599–606.

Kay DWK, Roth MC (1961) Environmental and hereditary factors in the schizophrenias of old age (late paraphrenia) and their bearing on the general problem of causation in schizophrenia. J Ment Sci 107:649–686.

Kendler KS (1983) Overview: A current perspective on twin studies of schizophrenia. Am J Psychiatry 140: 1413–1425.

Kendler KS (1989) Limitations of the ratio of concordance rates in monozygotic and dizygotic twins [letter]. Arch Gen Psychiatry 46:477–478.

Kendler KS, Robinette CD (1982) Schizophrenia in the National Academy of Sciences-National Research Council Twin Registry: a 16-year update. Am J Psychiatry 140:1551–1563.

Kendler KS, Pedersen NL, Farahmand BY, Persson P-G (1996) The treated incidence of psychotic and affective illness in twins compared with population expectation: a study in the Swedish Twin and Psychiatric Registries. Psychol Med 26:1135–1144.

Kinney DF, Jacobsen B (1978) Environmental factors in schizophrenia: new adoption study evidence. In: Wynne LC, Cromwell RL, Mattysse (eds.) The Nature of schizophrenia: new approaches to research and treatment. Wiley, New York, pp 38–51.

Kleist K (1925) Die gegenwärtigen Strömungen in der Psychiatrie. De Gruyter, Berlin Leipzig.

Kohl SG, Casey G (1975) Twin gestation. Mt Sinai J Med 42:523–539.

Kraepelin E (1909) Psychiatrie. 8. Auflage. Barth, Leipzig.

Kringlen E (1967) Heredity and environment in the functional psychoses. Universitets-Forlaget, Oslo.

Kringlen E (1971) Beiträge der neueren Zwillingsforschung zur Frage der Ätiologie und Pathogenese der Schizophrenie. In: Bleuler M, Angst J (Hrsg.) Die Entstehung der Schizophrenie. Huber, Bern.

Kringlen E (1990) Genetical aspects with emphasis on twin studies. In: Kringlen E, Lavik NJ, Torgersen S (eds.) Etiology of mental disorder. University of Oslo, Oslo.

Kringlen E, Cramer G (1989) Offspring of monozygotic twins discordant for schizophrenia. Arch Gen Psychiatry 46:873–877.

Krüger J, Propping P (1976) Rückgang der Zwillingsgeburten in Deutschland. Dtsch Med Wochenschr 101:475–480.

Kuban KCK, Leviton A (1994) Cerebral palsy. N Engl J Med 330:188–195.

Kulakowski S, Larroche JC (1980) Cranial computerized tomography in cerebral palsy. An attempt at anatomo-clinical and radiological correlations. Neuropediatrics 11:339–353.

Kurihara M (1959) A study of schizophrenia by the twin method. Psychiat Neurol Jap 61:1721–1741.

LaBuda MC, Gottesman II, Pauls DL (1993) Usefulness of twin studies for exploring the etiology of childhood and adolescent psychiatric disorders. Am J Med Genet 48:47–59.

Langman J (1977) Medizinische Embryologie. Deutsche Übersetzung von Drews U. Thieme, Stuttgart.

Leonhard K (1956) Aufteilung der endogenen Psychosen. Akademie, Berlin.

Leonhard K (1975) Ein dominanter und ein rezessiver Erbgang bei zwei verschiedenen Formen von Schizophrenie. Nervenarzt 46:242–248.

Leonhard K (1976) Genese der zykloiden Psychosen. Psychiatry Neurol Med Psychol 33:145–157.

Leonhard K (1978) Zwillingsuntersuchungen mit einer differenzierten Diagnose der endogenen Psychosen. Psychisch-soziale Einflüsse bei gewissen schizophrenen Formen. Psychiat Neurol Med Psychol 28:78–88.

Leonhard K (1979) Über erblich bedingte und psychosozial bedingte Schizophrenien. Psychiatry Neurol Med Psychol 31:606–626.

Leonhard K (1986) Different causative factors in different forms of schizophrenia. Br J Psychiatry 149:1–6.

Leonhard K (1995) Aufteilung der endogenen Psychosen und ihre differenzierte Ätiologie. 7. neubearbeitete und ergänzte Auflage. Thieme Verlag, Stuttgart New York.

Lewis SW, Chitkara B, Reveley AM, Murray RM (1987a) Family history and birthweight in monozygotic twins concordant and discordant for psychosis. Acta Genet Med Gemellol 36:267–273.

Lewis SW, Reveley AM, Reveley MA, Chitkara B, Murray RM (1987b) The familial/sporadic distinction as a strategy in schizophrenia research. Br J Psychiatry 151:306–313.

Lewis SW, Chitkara B, Reveley AM (1989) Hand preference in psychotic twins. Biol Psychiatry 25:215–221.

Little RE (1986) Birthweight and gestational age: Mother's estimates compared with state and hospital records. Am J Public Health 76:1350–1351.

Little J, Bryan E (1986) Congenital anomalies in twins. Semin Perinatol 10:50–64.

Loehlin JC, Nichols RC (1976) Heredity, environment, and personality. University of Texas Press, Austin.

Loranger AW (1984) Sex differences in age at onset of schizophrenia. Arch Gen Psychiatry 41:157–161.

Luchins D, Pollin W, Wyatt RJ (1980) Laterality in monozygotic schizophrenic twins: An alternative hypothesis. Biol Psychiatry 15:87–93.

Luxenburger H (1928) Vorläufiger Bericht über psychiatrische Serienuntersuchungen an Zwillingen. Z Ges Neurol Psychiatry 116:297–326.

Luxenburger H (1936) Untersuchungen an schizophrenen Zwillingen und ihren Geschwistern zur Prüfung der Realität von Manifestationserscheinungen. Z Ges Neurol Psychiatry 154:351–394.

Matheny AP (1979) Appraisal of parental bias in twin studies: ascribed zygosity and IQ differences in twins. Acta Genet Med Gemellol (Roma) 28:155–160.

Matheny AP, Wilson RS, Dolan AB (1976) Relations between twins' similarity of appearance and behavioral similarity: testing an assumption. Behav Genet 6:43–52.

McArthur N (1953) Statisticts in twin birth in Italy. 1949 and 1950. Ann Eugen 17:249.

McGue M (1992) When assessing twin concordance, use the probandwise not the pairwise rate. Schizophr Bull 18:171–176.

McGuffin P, Farmer AE, Gottesman II, Murray RM, Reveley AM (1984) Twin concordance for operationally defined schizophrenia: confirmation of familiality and heritability. Arch Gen Psychiatry 41:541–545.

McManus IC (1980) Handedness in twins: A critical review. Neuropsychologia 18:347–355.

McNeil TF (1987) Perinatal influences in the development of schizophrenics. In: Helmchen H, Henn FA (Hrsg) Biological perspectives of schizophrenia. Wiley, Chichester, pp 125–138.

McSweeney D, Timms P, Johnson A (1978) Thyro-endocrine pathology, obstetric morbidity and schizophrenia: a survey of a hundred families with a schizophrenic proband. Psychol Med 8:151–155.

Mednick SA, Machon RA, Huttunen MO, Bonett D (1988) Adult schizophrenia following prenatal exposure to an influenza epidemic. Arch Gen Psychiatry 45:189–192.

Meisami E, Firoozi M (1985) Acetylcholinesterase activity in the developing olfactory bulb: a biochemical study on normal maturation and the influence of peripheral and central connections. Dev Brain Res 21:115–124.

Modrewsky K (1980) The offspring of schizophrenic parents in a North Swedish isolate. Clin Genet 17:191–201.

Morison J (1949) Congenital malformations in one of monozygotic twins. Arch Dis Child 24:214–218.

Mosher LR, Pollin W, Stabenau JR (1971) Identical twins discordant for schizophrenia. Neurological findings. Arch Gen Psychiatry 24:422–4530.

Munsinger H, Douglass A (1976) The syntactic abilities of identical twins, fraternal twins, and their siblings. Child Dev 47:40–50.

Murphy D, Wyatt RJ (1972) Reduced monoamine oxidase activity in blood platelets from schizophrenic patients. Nature 238:225–226.

Murray RM, Murphy DL (1978) Drug response and psychiatric nosology. Psychol Med 8:667–681.

Nasrallah HA, Schwarzkopf SB, Coffman JA, Olson SC (1991) Developmental brain abnormalities on MRI in schizophrenia: the role of genetic and perinatal factors. In: Mednick SA, Cannon TD, Barr CE, Lyon M (eds.) Fetal and neural development and adult schizophrenia. Cambridge University Press, Cambridge, pp 216–223.

Nelson KB, Ellenberg JH (1986) Antecedents of cerebral palsy: multivariate analysis of risk. N Engl J Med 315:81–86.

Neumann H (1859) Lehrbuch der Psychiatrie. Enke, Erlangen.

Newell-Morris LL, Fahrenbruch CE, Sackett GP (1989) Prenatal psychological stress, dermatoglyphic asymmetry and pregnancy outcome in the pigtailed macaque. Biol Neonate 56:61–75.

Nicholas JW, Jenkins WJ, Marsh WL (1957) Human blood chimeras. A study of surviving twins. Br Med J I:1458.

Nichols R (1965) The national merit twin study. In: Vandenberg S (ed.) Methods and goals in human behavior genetics. New York.

Nimgaonkar VL, Wessely S, Murray RM (1988) Prevalence of familiality, obstetric complications, and structural brain damage in schizophrenic patients. Br J Psychiatry 153:191–197.

O'Callaghan E, Larkin C, Waddington JL (1990a) Obstetric complications in schizophrenia and the validity of maternal recall. Psychol Med 20:89–94.

O'Callaghan E, Larkin C, Kinsella A, Waddington JL (1990b) Obstetric complications, the putative familial-sporadic distinction, and tardive dyskinesia in schizophrenia. Br J Psychiatry 157:578–584.

O'Callaghan E, Sham P, Takei N, Glover G, Murray RM (1991a) Schizophrenia after prenatal exposure to 1957 A2 influenza epidemic. Lancet 337: 118–119.

O'Callaghan E, Gibson T, Colohan HA, Walshe D, Buckley P, Larkin C, Waddington JL. (1991b) Season of birth in schizophrenia. Evidence for confinement of an excess of winter births to patients without a family history of mental disorder. Br J Psychiatry 158:764–769.

O'Kusky JR (1985) Synapse elimination in the developing visual cortex: a morphometric analysis in normal and dark-reared cats. Dev Brain Res 22:81–91.

Oldfield RC (1971) The assessment and analysis of handedness: the Edinburgh Inventory. Neuropsychologia 9:97.

Onstad S, Skre I, Torgersen S, Kringlen E (1991a) Twin concordance for DSM III-R schizophrenia. Acta Psychiatry Scand 83:395–401.

Onstad S, Skre I, Edvardsen J, Torgersen S, Kringlen E (1991b) Mental disorders in first-degree relatives of schizophrenics. Acta Psychiatry Scand 83:463–467.

Onstad S, Skre I, Torgersen S, Kringlen E (1992) Birthweight and obstetric complications in schizophrenic twins. Acta Psychiatry Scand 85:70–73.

Owen MJ, Lewis SH, Murray RM (1988) Obstetric complications and schizophrenia: a computed tomographic study. Psychol Med 18:331–339.

Owens DGC, Johnstone EC, Crow TJ, Frith CD, Jagoe JR, Kreel L (1985) Lateral ventricular size in schizophrenia: relationship to the disease process and its clinical manifestations. Psychol Med 15:27–41.

Oxenstierna G, Bergstrand G, Bjerkenstedt L, Sedvall G, Wik G (1984) Evidence of disturbed CSF circulation and brain atrophy in cases of schizophrenic psychosis. Br J Psychiatry 144:654–661.

Parnas J, Schulsinger F, Teasdale TW, Schulsinger H, Feldman PM, Mednick SA (1982) Perinatal complications and clinical outcome within the schizophrenic spectrum. Br J Psychiatry 140:416–420.

Parnas J (1991) Schizophrenia: etiological factors in the light of longitudinal high-risk research. In: Kringlen E, Lavik NJ, Torgersen S (eds.) etiology of mental disorder. University of Oslo, Oslo, pp 49–61.

Pearlson GD, Garbacz DJ, Moberg PJ, Ahn HS, dePaulo JR (1985) Symptomatic, familial, perinatal, and social correlates of computerized axial tomography changes in schizophrenics and bipolars. J Nerv Ment Dis 173:42–50.

Perris C (1974) A Study of cycloid psychosis. Acta Psychiatr Scand [Suppl] 253.

Pfuhlmann B, Franzek E, Stöber G, Cetkovich-Bakmas M, Beckmann H (1997) On interrater reliability for Leonhard's classification of endogenous psychoses. Psychopathology 30:100–105.

Pfuhlmann B, Stöber G, Franzek E, Beckmann H (1998) Cycloid psychoses predominate in severe postpartum psychiatric disorders. J Affect Disord 50:125–134.

Plomin R, Willerman L, Loehlin JC (1976) Resemblance in appearance and the equal environments assumption in twin studies of personality traits. Behav Genet 6:43–52.

Pollin W, Stabenau JR, Tupin J (1965) Family studies with identical twins discordant for schizophrenia. Psychiatry 28:60–76.

Pollin W, Stabenau JR, Mosher L, Tupin J (1966) Life history differences in identical twins discordant for schizophrenia. Am J Orthopsychiatry, 36:492–509.

Pollin W, Allen M, Hoffer A (1969) Psychopathology in 15909 pairs of veteran twins: evidence for a genetic factor in the pathogenesis of schizophrenia and its relative absence in psychoneurosis. Am J Psychiatry 126:597–610.

Pollin W (1972) The pathogenesis of schizophrenia. Arch Gen Psychiatry 27:29–37.

Poltorak M, Wright R, Hemperly JJ, Torrey EF, Issa F, Wyatt RJ, Freed WJ (1997) Monozygotic twins discordant for schizophrenia are discordant for N Cam and L1 in CSF. Brain Res 751:152–154.

Pope HG, Lipinsky JF (1978) Diagnosis in schizophrenia and manic-depressive illness. Arch Gen Psychiatry 35:811–828.

Propping P (1983) Zwillingsforschung. In: Autrum H, Wolf U (Hrsg.) Humanbiologie. Springer, Berlin Heidelberg New York Tokyo, pp 143–153.

Propping P (1984) Norm und Variabilität – Der Krankheitsbegriff in der Genetik. Universitas 39:1271–1281.

Propping P (1989) Psychiatrische Genetik. Befunde und Konzepte. Springer, Berlin Heidelberg New York Tokyo.

Propping P, Krüger J (1976) Über die Häufigkeit von Zwillingsgeburten. Dtsch Med Wochenschr 101:506–512.

Price D (1950) Primary biases in twin studies: a review of prenatal and natal difference – producing factors in monozygotic pairs. Am J Hum Gent 2:293.

Putten DMV, Torrey EF, Larive AB, Merril CR (1996) Plasma Protein Variations in monozygotic twins discordant for schizophrenia. Biol Psychiatry 40:437–442.

Quitkin F, Rifkin A, Tsuang MT, Kane JM, Klein DF (1980) Can schizophrenia with premorbid asociality be genetically distinguished from the other form of schizophrenia? Psychiatry Res 2:99–105.

Rakic P (1988) Specification of cerebral areas. Science 421:170–176.

Rausen AR, Seki M, Strauss L (1965) Twin transfusion syndrome. J Pediatr 66:613–628.

Reddy R, Mukherjee S, Schnur DB, Chin J, Degreef G (1990) History of obstetric complications, family history, and CT-scan findings in schizophrenic patients. Schizophr Res 3:311–314.

Reveley AM, Reveley MA, Clifford CA, Murray RM (1982) Cerebral ventricular size in twins discordant for schizophrenia. Lancet i:540–541.

Reveley MA, Reveley AM, Clifford CA, Murray RM (1983) Genetics of platelet MAO activity in discordant schizophrenic and normal twins. Br J Psychiatry 142:560–565.

Reveley AM, Reveley MA, Murray RM (1983) Enlargement of cerebral ventricles in schizophrenics is confined to those without known genetic predisposition [letter]. Lancet 2:525.

Reveley AM, Reveley MA, Murray RM (1984) Cerebral ventricular enlargement in non-genetic-schizophrenia: a controlled twin study. Br J Psychiatry 144:89–93.

Rieder RO, Rosenthal D, Wender P, Blumenthal H (1975) The offspring of schizophrenics. Arch Gen Psychiatry 32:200–211.

Rosanoff AJ, Handy LM, Plesset IR, Brush S (1934) The etiology of so-called schizophrenic psychosis. With special reference to their occurrence in twins. Am J Psychiatry 91:247–286.

Rosenthal D (1959) Some factors associated with concordance and discordance with respect to schizophrenia in monozygotic twins. J Nerv Ment Dis 129:1–11.

Rosenthal D (1963) The Genain quadruplets. Basic Books, New York.

Sarna S, Kaprio J (1980) Optimization prozedures in twin zygosity by markers. Acta Genet Med Gemellol 28:139–148.

Scarr S (1968) Environmental bias in twin studies. In: Vandenberg S (ed.) Progress in human behavior genetics. Hopkins University Press, Baltimore.

Scarr S, Carter-Saltzman L (1979) Twin method: defence of a critical assumption. Behav Genet 9:527–542.

Schaumann B, Alter M (1976) Dermatoglyphics in medical disorders. Springer, New York.

Schepank H (1974) Erb- und Umweltfaktoren bei Neurosen. Tiefenpsychologische Untersuchungen an 50 Zwillingspaaren. Springer, Berlin Heidelberg New York.

Schepank H (1993) Why are monozygotic twins so different in personality? In: Bouchard TJjr., Propping P (eds.) Twins as a tool of behavioral genetics. Wiley, Winichester, pp 139–146.

Schmidt L (1986) Paarbeziehung und Persönlichkeitsentwicklung von Zwillingen. In: Friedrich W, Kabat vel Job O (Hrsg.) Zwillingsforschung international. Verlag der Wissenschaften, Berlin.

Schwarzkopf SB, Nasrallah HA, Olson SC, Coffman JA, McLaughlin JA (1989) Perinatal complications and genetic loading in schizophrenia: preliminary findings. Psychiatry Res 27:233–239.

Seeman MV (1982) Gender differences in schizophrenia. Can J Psychiatry 27:107–112.

Sham P, O'Callaghan E, Takei N, Murray GK, Hare EH, Murray RM (1992) Schizophrenia following prenatal exposure to influenza epidemics between 1939 and 1960. Br J Psychiatry 160:451–466.

Shields J (1962) Monozygotic twins. Oxford University Press, London.

Shields J, Gottesman II (1972) Cross-National diagnosis of schizophrenia in twins. Arch Gen Psychiatry 27:725–730.

Shields J, Gottesman II, Slater E (1967) Kallmann's 1946 schizophrenic twin study in the light of new information. Acta Psychiatry Scand 43:385–396.

Siemens HW (1924) Die Zwillingspathologie. Ihre Bedeutung. Ihre Methodik. Ihre bisherigen Ergebnisse. Springer, Berlin Heidelberg New York Tokyo.

Slater E (1953) Psychotic and neurotic illness in twins. Med Res Counc Spect Dept Ser No 278. Her Majesty's Stationary Office, London.

Slater E, Cowie V (1971) The genetics of mental disorders. Oxford University Press, London.

Sobel DE (1961) Infant mortality and malformations in children of schizophrenic women. Psychiatry Q 35:60–65.

Spellacy WN (1988) Antepartum complications in twin pregnancies. In: Gall SA (ed.) Clinics in perinatology. Saunders, Philadelphia.

Spitz RA (1945) Hospitalism: an inquiry into the genesis of psychiatric conditions in early childhood. Psychoanal Study Child 1:53–74.

Spitzer RL, Williams JBW (1984) Structured clinical interview for DSM-III (SCID), 5/1/84 revision. Biometrics Research Department, York State Psychiatric Institute, New York.

Stabenau JR, Pollin W (1967) Early characteristics of monozygotic twins discordant for schizophrenia. Arch Gen Psychiatry 17:723–734.

Stevenson AC, Johnston HA, Stewert MIP, Golding DR (1966) Congenital malformations. A report of a study of series of consecutive births in 24 centres. Bull Wld Hlth Org [Suppl] 34.

Stöber G, Franzek E, Beckmann H (1992) The role of maternal infectious diseases during pregnancy in the etiology of schizophrenia in the offspring. Eur Psychiatry 7:147–152.

Stöber G, Franzek E, Beckmann H (1993a) Schwangerschafts- und Geburtskomplikationen – ihr Stellenwert in der Entstehung schizophrener Psychosen. Fortschr Neurol Psychiat 61:329–337.

Stöber G, Franzek E, Beckmann H (1993b) Obstetric complications in distinct schizophrenic subgroups. Eur Psychiatry 8:293–299.

Stöber G, Franzek E, Lesch KP, Beckmann H (1995) Periodic catatonia: a schizophrenic subtype with major gene effect and anticipation. Eur Arch Psychiatry Clin Neurosci 245:135–141.

Strauss JS, Kokes RF, Klorman R, Sacksteder JL (1977) Premorbid adjustment in schizophrenia: concepts, measures, and implications. Part I: the concept of premorbid adjustment. Schizophr Bull 3:182–185.

Strik WK, Dierks T, Franzek E, Maurer K, Beckmann H (1993) Differences in P300-amplitude and topography between cycloid psychosis and schizophrenia in Leonhard's classification. Acta Psychiatry Scand 87:179–183.

Strik WK, Fallgatter AJ, Stöber G, Franzek E, Beckmann H (1996) Spezific P300 features in patients with cycloid psychosis. Acta Psychiatry Scand 94:471–476.

Strömgren, E (1936) Zum Ersatz des Weinbergschen "abgekürzten Verfahrens". Zeitschr ges Neurol Psychiatrie 153:784–797.

Suddath RL, Christison GW, Torrey EF, Casanova MF, Weinberger DR (1990) Anatomical abnormalities in the brains of monozygotic twins discordant for schizophrenia. N Engl J Med 322:789–794.

Sydow G, Rinne A (1958) Very unequal "identical twins". Acta paediat 47:163–171.

Tienari P (1963) Psychiatric illness in identical twins. Acta Psychiatry Scand [Suppl] 171:1–195.

Tienari P (1968) Schizophrenia in monozygotic male twins. In: Rosenthal D, Kety SS (eds.) The Transmission of schizophrenia. Pergamon Press, New York.

Torgersen S (1979) The determination of twin zygosity by means of mailed questionaire. Acta Genet Med Gemellol 28:225–236.

Torrey EF, Rawlings R, Waldman IN (1988) Schizophrenic births and viral diseases in two states. Schizophr Res 1:73–77.

Torrey EF, Ragland JD, Gold JM, Goldberg TE, Bowler AE, Bigelow LB, Gottesman II (1993a) Handedness in twins with schizophrenia: was Boklage correct? Schizophr Res 9:83–85.

Torrey EF, Bowler AE, Rawlings R, Terrazas A (1993b) Seasonality of schizophrenia and stillbirths. Schizophr Bull 19:557–562.

Torrey EF, Rawlings RR, Ennis JM, Merrill DD, Flores DS (1996) Birth seasonality in bipolar disorder, schizophrenia, schizoaffective disorder and stillbirths. Schizophr Res 21:141–149.

Tsuang MT, Fowler RC, Cadoret RJ, Monelly E (1974) Schizophrenia among first degree relatives of paranoid and non paranoid schizophrenics. Compr Psychiatry 15:295–301.

Turner SW, Toone KB, Brett-Jones JR (1986) Computerized tomographic scan changes in early schizophrenia – preliminary findings. Psychol Med 16:219–225.

Vandenberg SG, Wilson K (1979) Failure of the twin situation to influence twin differences in cognition. Behav Genet 9:55–60.

Van Os J, Fañanas L, Cannon M, Macdonald A, Murray R (1997) Dermatoglyphic abnormalities in psychosis: a twin study. Biol Psychiatry 41: 624–626.

Videbech TH, Weeke A, Dupont A (1974) Endogenous psychoses and season of birth. Acta Psychiatry Scand 50:202–218.

Vogel F, Motulsky AG (1986) Human genetics. Springer, Berlin Heidelberg New York.

Wakita Y, Narahara K, Kimoto H (1988) Multivariate analysis of dermatoglyphics of severe mental retardates: an application of the constellation graphical method for discriminant analysis. Acta Med Okayama 42:159–168.

Weinberg W (1902) Beiträge zur Physiologie und Pathologie der Mehrlingsgeburten beim Menschen und Probleme der Mehrlingsgeburtenstatistik. Z Geburtshilfe Gynäkol 47:12.

Weinberg W (1909) Der Einfluß von Alter und Geburtenzahl der Mutter auf die Häufigkeit der ein- und zweieiigen Zwillingsgeburten. Z Geburtshilfe Gynäkol 65:318–324.

Weinberger DR (1987) Implications of normal brain development for the pathogenesis of schizophrenia. Arch Gen Psychiatry 44:660–669.

Weinberger DR, Berman KF, Torrey EF (1992a) Correlations between abnormal hippocampal morphology and prefrontal physiology in schizophrenia. Clin Neuropharmacol 15 [Suppl] 1 pt A 393A–394A.

Weinberger DR, Berman KF, Suddath R, Torrey EF (1992b) Evidence of dysfunction of a prefrontal-limbic network in schizophrenia: a magnetic resonance imaging and regional cerebral blood flow study of discordant monozygotic twins. Am J Psychiatry 149:890–897.

Weltgesundheitsorganisation (1991) Internationale Klassifikation psychischer Störungen: Dilling H, Mobour W, Schmidt MH (Hrsg.) ICD 10. Huber, Bern.

Wenar C, Coulter JB (1962) A reliability study of developmental histories. Child Development 33:453–462.

Wernicke C (1900) Grundriß der Psychiatrie in klinischen Vorlesungen. Barth, Leipzig.

Wilcox J, Nasrallah HA (1987) Perinatal distress and prognosis of psychotic illness. Neuropsychobiology 17:173–175.

Wyatt RJ, Murphy D, Belmaker R (1973a) Reduced monoamine oxidase activity in platelets: a possible genetic marker for vulnerability to schizophrenia. Science 173:916–918.

Wyatt RJ, Saavedra JM, Belmaker R, Cohen S, Pollin W (1973b) The dimethyl-tryptamine-forming enzyme in blood platelets: A study in monozygotic twins discordant for schizophrenia. Am J Psychiatry 130:1359–1361.

Zerbin-Rüdin E (1974) Vererbung und Umwelt bei der Entstehung psychischer Störungen. Wissenschaftliche Buchgesellschaft, Darmstadt.

Zerbin-Rüdin E (1980) Gegenwärtiger Stand der Zwillings- und Adoptionsstudien zur Schizophrenie. Nervenarzt 51:379–391.

Zipursky RB, Schulz SC (1987) Seasonality of birth and CT findings in schizophrenia. Biol Psychiatry 22:1288–1292.

Index

Acknowledgements

We thank Prof. Dr. A. Schmidtke and Prof. Dr. W. Strik for the statistical advice and evaluation of the data.

We thank Prof. Dr. G. Jungkunz, Director of the Psychiatric Hospital Lohr/Main and Dr. Schottky, Director of the Psychiatric Hospital Werneck, for permission to look through their clinics' medical records of the twin subjects and for their excellent co-operation in this study.

We thank the Deutsche Forschungsgemeinschaft for their generous support, without which this study would not have been possible.

We thank the twin subjects and their families for their willingness to take part in this study.